Najeh TKA

New route to the synthesis of chiral 1,2,4-oxadiazin-5-ones

Najeh TKA

New route to the synthesis of chiral 1,2,4-oxadiazin-5-ones

easy synthesis of new chiral 1,2,4-oxadiazin-5-ones from natural amino acids

ScienciaScripts

Imprint

Cover image: www.ingimage.com

This book is a translation from the original published under ISBN 978-620-6-70703-5.

Publisher:
Sciencia Scripts
is a trademark of
Dodo Books Indian Ocean Ltd. and OmniScriptum S.R.L publishing group

120 High Road, East Finchley, London, N2 9ED, United Kingdom
Str. Armeneasca 28/1, office 1, Chisinau MD-2012, Republic of Moldova, Europe
Printed at: see last page
ISBN: 978-620-7-92433-2

Contents

GENERAL INTRODUCTION

The living world is made up of asymmetric structures synthesised by highly selective processes occurring in organisms. Natural chiral molecules often exist in the form of a single enantiomer, such as amino acids, which are the building blocks of proteins and always have the (*L*) configuration.

Since the discovery of chirality, chemists have devoted all their creativity to developing effective systems for controlling chirality. In 1848, Louis Pasteur discovered the notion of chirality[1] while working on the molecule tartaric acid **1 (Diagram 1)**. The work of French chemist Joseph Le Bel and Dutch chemist Van't Hoff followed to explain the concept of configuration, which stems directly from stereochemistry.[2]

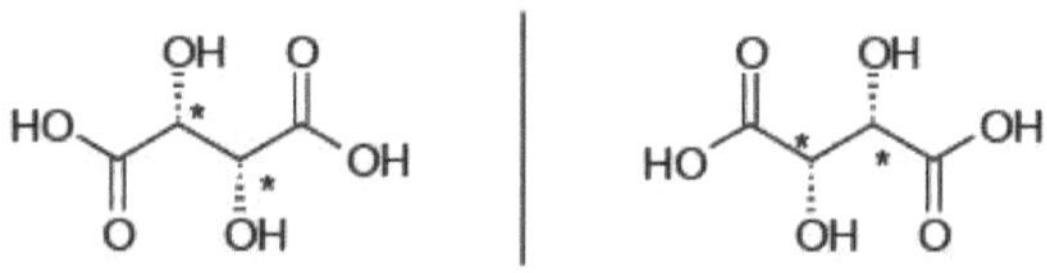

(2*R*, 3*S*)-(+)-acide tartrique | **(2*S*, 3*R*)-(-)-acide tartrique**

1 | **1'**

Schéma 1

A chiral molecule in one or other enantiomeric form will not have the same effect on our body, as is the case for certain drugs. The first recognised example illustrating the different activities of two enantiomers was described in 1849 by Puitti, who noted that the (*L*)-asparagine **2** amide of aspartic acid, a natural amino acid, is tasteless (no taste), whereas its (*D*)-enantiomer has a sweet taste[3,4] **(Diagram 2)**.

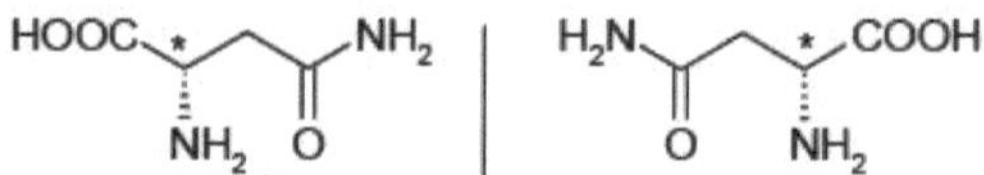

2 | **2'**

(*L*)-aspargine | **(*D*)-aspargine**

Insipide | **Goût sucré**

Schéma 2

For example, (*R*)-limonene **3**, used in the perfume industry, has an orange odour, while (*S*)-limonene has a lemon odour.

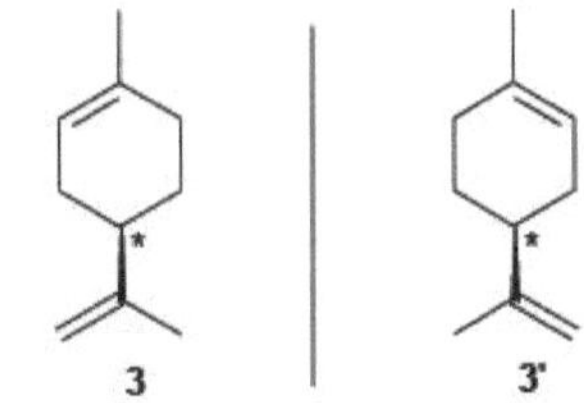

3 | 3'

(*R*)-Limonène | (*S*)-Limonène

Odeur orange | Odeur citron

Schéma 3

Nowadays, the molecules used in pharmaceuticals are increasingly developed in the form of a single enantiomer. The main reason for this is that one of the two enantiomers has the desired property, while the other enantiomer is toxic or simply inactive. One of the saddest examples of the use of racemic mixtures is the case of thalidomide[4] **4**. This drug was mainly used by pregnant women in the early months of pregnancy to treat nausea. While the (*R*) form is an excellent analgesic, the (*S*) form is a poison that causes *foetal* malformations **(Diagram 4).**

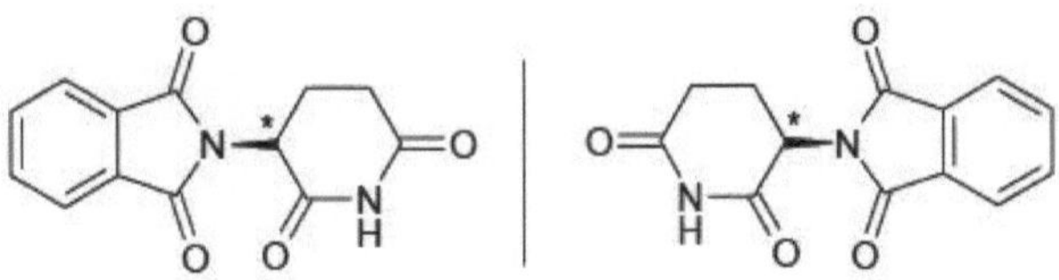

4 | 4'

(*S*)-Thalidomide | (*R*)-Thalidomide

tératogène | analgésique

Schéma 4

Another example is Ketamine **5**, the (*S*)-form of which is used as an anaesthetic and the (*R*)-form of which can cause psychological disorders[5] **(Diagram 5)**.

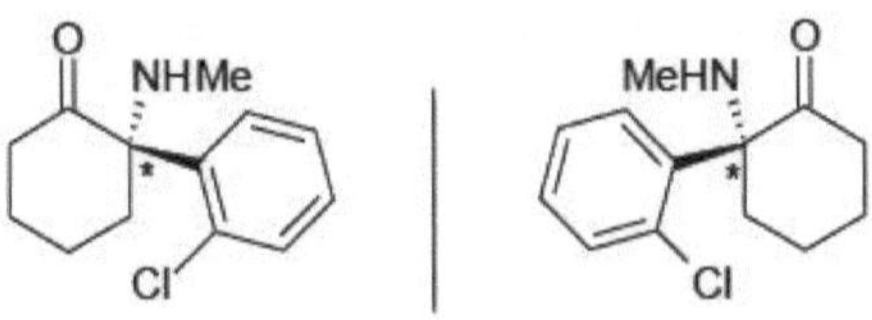

5 | 5'

(*S*)-Kétamine | (*R*)-Kétamine

anesthésique | excitant

Schéma 5

This constraint clearly demonstrates the importance of strict control of stereochemistry. For these reasons, the FDA (Food and Drug Administration) is encouraging pharmaceutical companies to market their drugs in the form of a single enantiomer. As a result, chemists have been urged to find new techniques for obtaining chiral compounds in optically pure form. Today, there are three main techniques:

- Splitting the racemic mixture.
- Asymmetric synthesis using a prochiral substrate.
- The use of a chiral synthon.

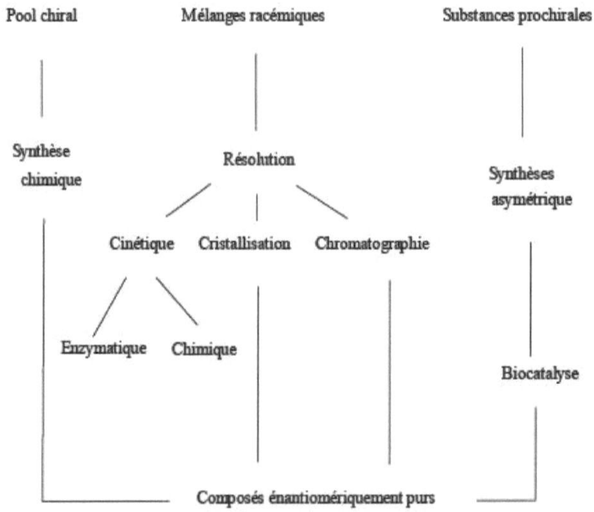

Chiral pool Racemic mixtures Proxy substances
Chemical synthesis
Resolution
Asymmetric synthesis
Kinetics Crystallization Chromatography
Enzymatic Chemical
Biocatalysis
Enantiomerically pure compounds

Fig.1*:* The main techniques for accessing enantiopure molecules

Splitting the racemic mixture involves separating the two enantiomers obtained in the form of a racemic mixture using a chiral auxiliary. Two diastereomers are formed whose properties are sufficiently different for them to be separated by conventional methods such as crystallisation and chromatography. The final variant of resolution involves the selective transformation of one of the two enantiomers: this is chemical or enzymatic kinetic resolution.

Asymmetric synthesis involves introducing an asymmetric element into a prochiral species using a chiral reagent or a chiral catalyst.

The use of a chiral synthon consists of carrying out a synthesis from optically pure products (chiral pool) while maintaining the chirality throughout the synthesis. Chiral synthons of natural origin include optically pure a-amino acids, which are also used as starting materials to prepare a wide range of natural products with pharmacological properties[7].

The research work carried out in this dissertation is linked to the last method and aims to develop a simple and rapid synthesis route for new nitrogenous and oxygenated heterocycles known for their interesting biological activities[8] .

The present work is part of our laboratory's ongoing work on the synthesis of various chiral heterocycles from a-amino acids, such as: chiral 1,3-oxazolidin-2-ones[9] ; chiral 1,2-benzothiazin-3-ones[9] ; saccharin analogues[10] ; 1,2,3-benzothiadiazines[11] and 1,2,4-oxadiazoles[12] .

This work was devoted to the synthesis of new chiral 1,2,4-oxadiazin-5-ones from natural amino acids. It is divided into three parts:

- The first part deals with the synthesis of a series of optically active a-bromoacids by nitrous deamination of natural a-amino acids.

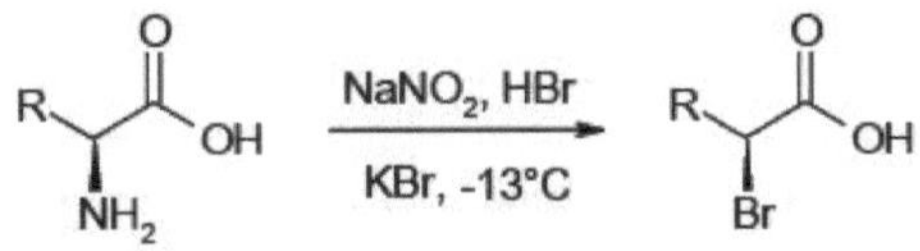

a : R=Me; **b** : R=i-Pr; **c** : R=i-Bu; **d** : R=Sec-Bu; **e** : R=Bn; **f*** : R=i-Bu

- The second part describes the synthesis of optically active a-bromoesters by esterification of the a-bromo acids obtained previously.

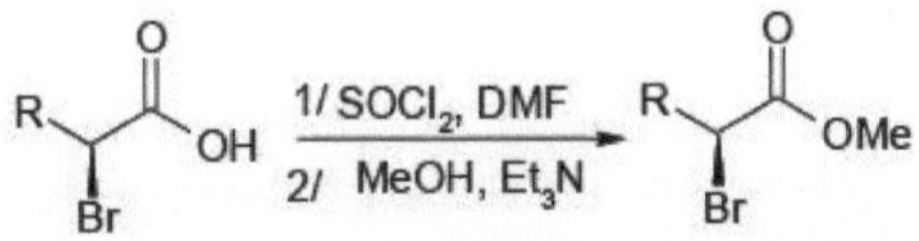

a : R=Me; **b** : R=i-Pr; **c** : R=i-Bu; **d** : R=Sec-Bu; **e** : R=Bn; **f*** : R=i-Bu

- The third part is devoted to the synthesis of new 1,2,4-oxadiazin-5-ones by condensation of the a-bromoesters obtained with (*25*)-1-benzenesulfonylpyrrolidine-2- carboxamidoxime prepared in our laboratory from proline[13] .

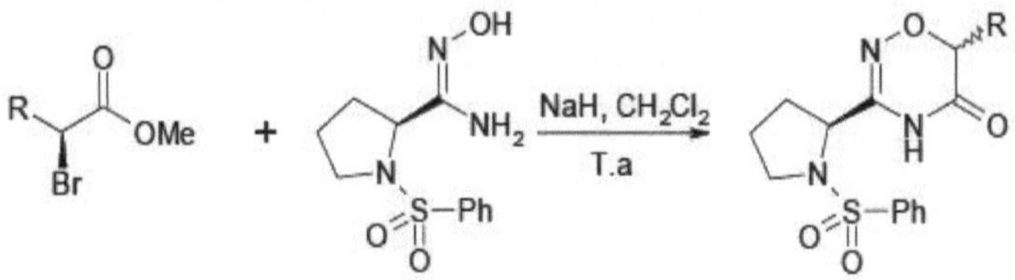

a : R=Me; **b** : R=i-Pr; **c** : R=i-Bu; **d** : R=Sec-Bu; **e** : R=Bn

Chapter 1

Synthesis of optically active a'-brominated acids by nitrous deamination of natural amino acids

A. General information and bibliography

Chiral a-bromo acids are widely described in the literature as key intermediates in organic synthesis and are often used as starting materials in the synthesis of biologically active compounds[16-29] . Thanks to the high mobility of bromine, a-bromoacids are ideal substrates for nucleophilic substitutions. A-bromoacids can be easily converted into a large number of products of synthetic interest by nucleophilic substitution with various amines, alcohols and sulphur products[14] . In particular, chiral a-bromoacids are sought after in the field of asymmetric synthesis[14-15] and several teams have attempted to prepare them in enantiomerically pure form.

In this chapter, we will first give an overview of the different methods of synthesising a-bromo acids and their applications. We will then present our own work, which consists of preparing optically active a-bromoacids by nitrous deamination of natural amino acids.

I. Synthetic benefits of brominated acids

A-bromoacids have found applications in various areas of organic synthesis. Here are a few examples:

I.1 Synthesis of *N-methylamino acids* from a-bromoacids

N-methylamino acids are interesting chemicals in the pharmaceutical field[16] . They are obtained simply by a method developed by Fisher and Michel[17] which involves reacting optically active a-bromo acids **6** with an excess of methylamine at 0°C. This method enabled the preparation of N-methylalanines **7a**, N-methylucines **7b** and N-methylphenylalanines **7c** in excellent chemical yields according to the following reaction scheme **(Scheme 8)**.

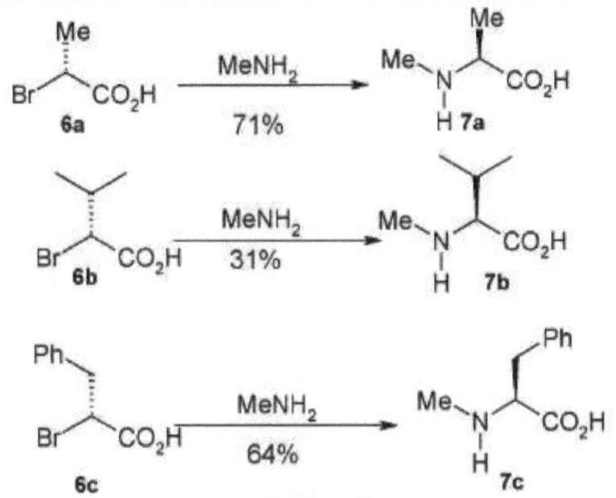

Schéma 8

This method[18] has also been used to prepare (D)-Surinamine **10** from a-amino acid **8** via a-bromo acid **9**.

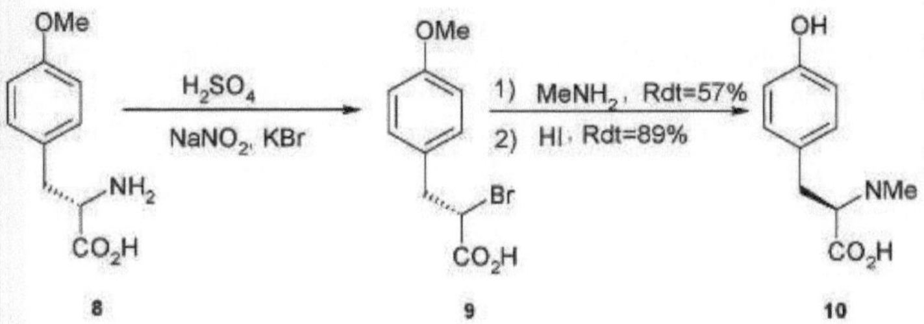

Schéma 9

I-2. Synthesis of optically active a-bromonitriles from a-bromoacids

In recent years, our laboratory has developed two methods for preparing optically active a-bromonitriles **12** from a-bromo acids. The a-bromonitriles were prepared by two different routes: The first involves reacting a-bromoacids **11** with chlorosulphonyl isocyanate (CSI) in the presence of trimethylamine. The second method involves first treating the a-bromo acids with thionyl chloride and ammonia to give a-bromoamides **13** which are then dehydrated either with phosphoryl chloride ($POCl_3$) or with thionyl chloride.

(Schéma 10).

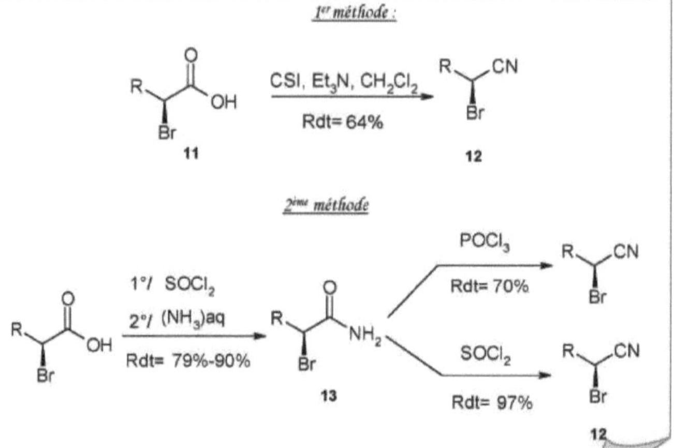

Schéma 10

I-3. Synthesis of a-hydrazino acids and their derivatives from a-bromo acids

A-hydrazino acids have antiviral and antitumour biological activities[20] . Malin et al,[21] synthesised a-hydrazino acids **15** from a-bromo acids **14** by condensation of α-bromo acids with excess hydrazine in ethanol at 70°C for 6-12 hours. The hydrazino acids **15** were obtained in chemical yields ranging from 58% to 71% **(Scheme 11).**

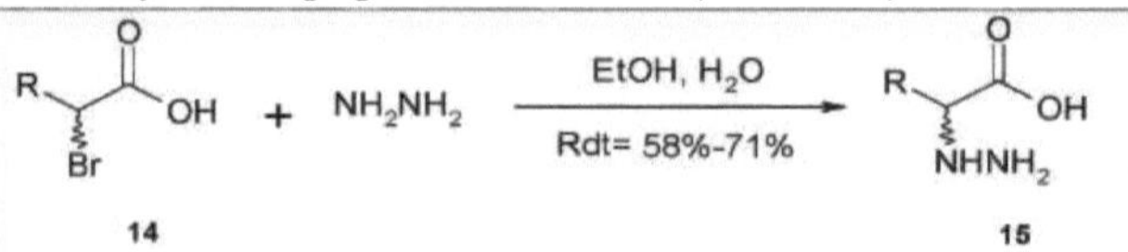

a: R=Me; b: R=Et; c:R=i-Pr; d: R=Ph

Schéma 11

I-4. Synthesis of levetiracetam from a-bromoacids

Type **20** levetiracetam has long attracted the interest of researchers because of its antiepileptic property[22] . Francesca et al,[23] described a simple method for obtaining levetiracetam from optically pure (*R*)-α-bromobutanoic acid **18**. The latter is obtained by resolution of a racemic mixture of two enantiomers of a-bromobutanoic acid **16** using a chiral alcoholic auxiliary of type **17**. Condensation of (*R*)-2-bromobutanoic acid **18** with 2-pyrolidinone **19** gives (S)-2-(2-oxopyrrolin-1-yl)butanoic acid which is subsequently converted to the desired (2S)-2-(2-oxopyrrolidinone)butamide **20** as shown in the following reaction scheme:

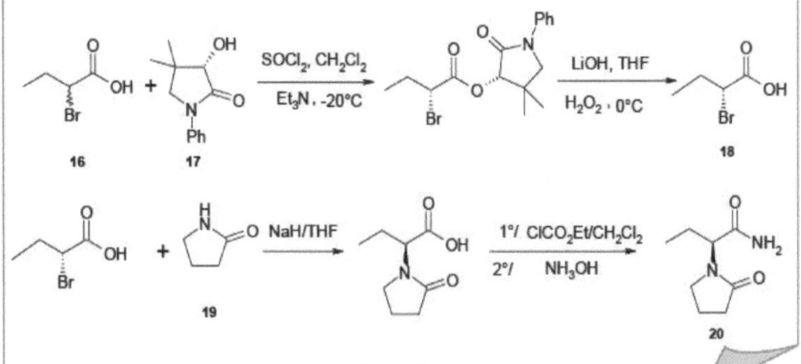

Schéma 12

I-5. Synthesis of amino acids using a halo acid

The first synthesis of glycine was reported by Cahours[24] in 1858. The addition of chloroacetic acid **21** to an ammoniacal solution of ethanol leads to the formation of the amino acid **22a** and two secondary products, the corresponding secondary amines **22b** and tertiary amines **22c**[25] **(Scheme 13).**

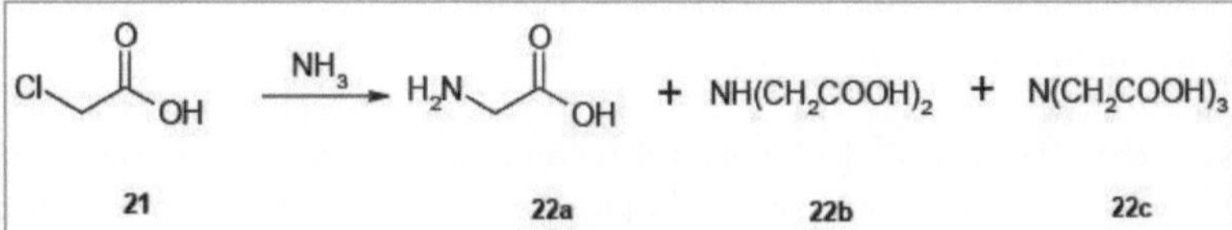

Schéma 13

In order to improve this reaction, tests have shown that a-amino acid **23** preferentially reacts with a-bromo acid for four days at room temperature using an alcoholic ammonia solution.[26] **(Diagram 14).**

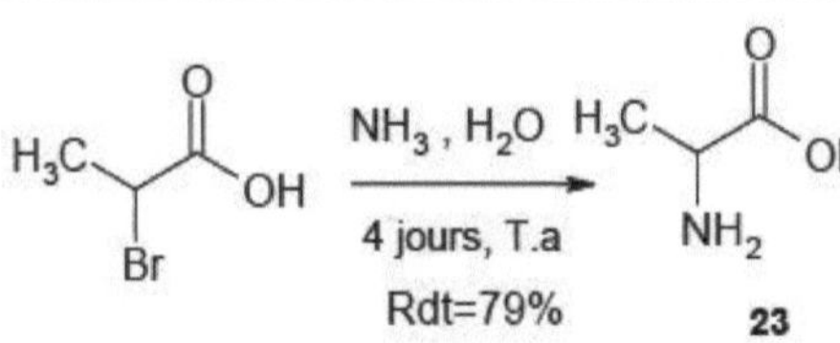

Schéma 14

I-6. Synthesis of a-hydroxyacids from a-bromoacids

A-hydroxy acids are widely described in the literature as important intermediates in asymmetric synthesis[27] and as compounds with interesting biological and pharmacological properties[28] . Horst et al,[29] described the synthesis of a-hydroxy acids **28** from a-bromoacids in four steps. The a-bromo acids **24** are first esterified with 2-methylpropenes **25** in an acid medium to give the corresponding a-bromoesters **26**. Treatment of the latter with caesium p-nitrobenzoate **27** leads to the compounds **28**. The a-hydroxy acids **29** are obtained by basic hydrolysis of compound 28 followed by acid saponification in dichloromethane **(Scheme 15)**.

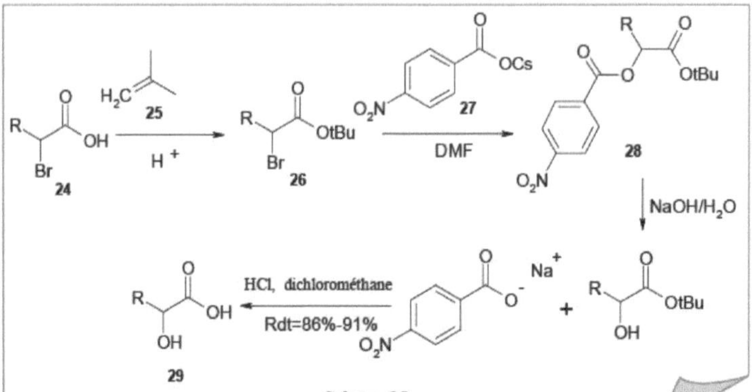

Schéma 15

II. Main methods for synthesising a-bromo acids

The methods for preparing a-bromoacids are very varied. In the following we describe the main methods cited in the literature.

II-1. Synthesis of a-bromoacids from carboxylic acids

II-1.a. Use of halophosphorus reagents

The synthesis of a-bromoacids from the corresponding carboxylic acids using a Hell-Volhard-Zelinsky (H-V-Z) reaction[30] has been widely described. In the most recent cases, a-bromination of the acids is carried out in the presence of a trivalent halophosphorus reagent such as PBr3 and PCl3 or a pentavalent reagent such as PBr5 and PCl5[31] .

From a mechanistic point of view, PBr_3 allows the formation of acid bromide 31. The enolic form 32 of the acid bromide then reacts with bromine to give the a-bromo acid bromide **33**, which is hydrolysed to give the a-bromo acids 34.

(Schéma 16).

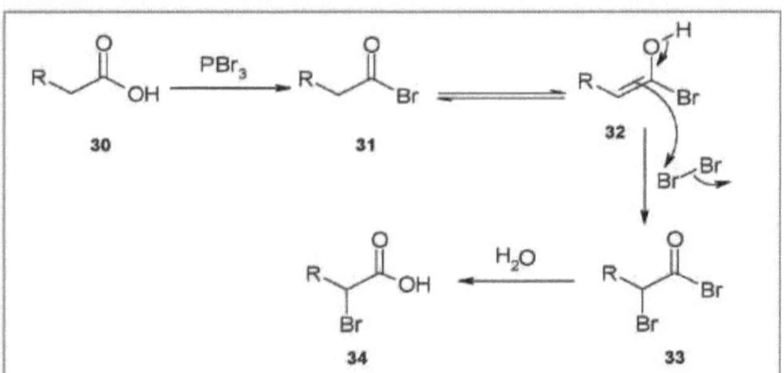

Schéma 16

II-1.b Use of N-bromosuccinimides

Lino et al,[32] have developed a new, simple and effective method for the bromination of carboxylic acids. This method involves treating carboxylic acids 36 with 1.5 equivalents of *N-bromosuccinimide* (NBS) in trifluoroacetic acid in the presence of a catalytic amount of sulphuric acid. The a-bromoacids 37 were obtained after 16 hours at a temperature of 85°C in excellent chemical yields of between 82% and 89% **(Scheme 17)**.

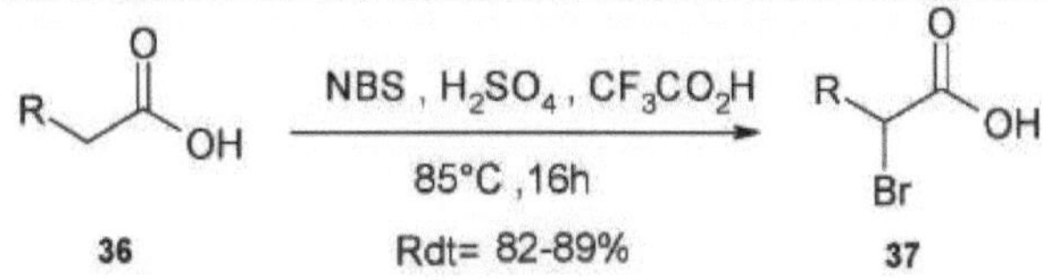

Schéma 17

II-2. Synthesis of α-bromoacids from β-ketoesters

Philip et al,[33] described the four-step synthesis of α-bromoacids from β-ketoesters. The first step involves condensation of an alkyl halide with β-ketoester 38 in a basic medium leading to the substituted β-ketoester of type **39**. This intermediate then undergoes carbonyl a-bromination to give the corresponding a-bromo-ß-ketoester **40**. This is then converted directly to the a-bromoester of type 41 by deacetylation in the presence of anhydrous barium hydroxide at 0°C. Finally, the a-bromo acid 42 is obtained by reflux heating in anhydrous benzene containing a catalytic amount of p-toluenesulphonic acid **(Scheme 18)**.

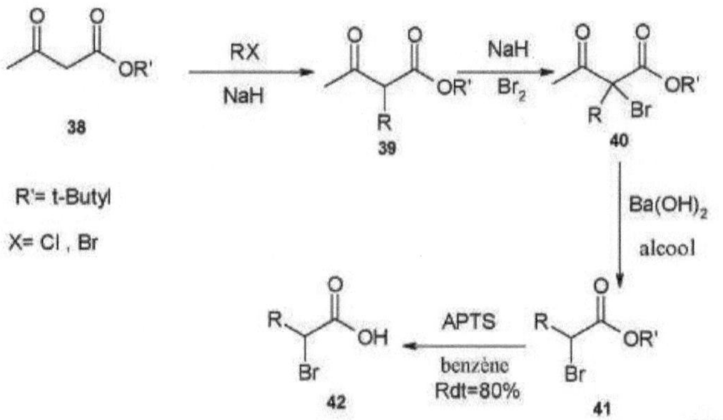

Schéma 18

II-3. Synthesis of a-bromoacids from F-alkyl ethylene

F-alkyl-a-bromoacids 46 were obtained from F-alkylethylenes 43 using a three-step procedure described by Baklouti's team.[34] Ethylenic derivatives of type 43 are commercial products, and their bromination is carried out in a reaction with acetoxymercury to give the brominated acetate 44. Hydrolysis of the latter under saponification conditions gives the brominated alcohol **45**. A Jones oxidation reaction leads to the a-brominated acid of type 46 in chemical yields of supérieurs à 90% **(Schéma 19)**.

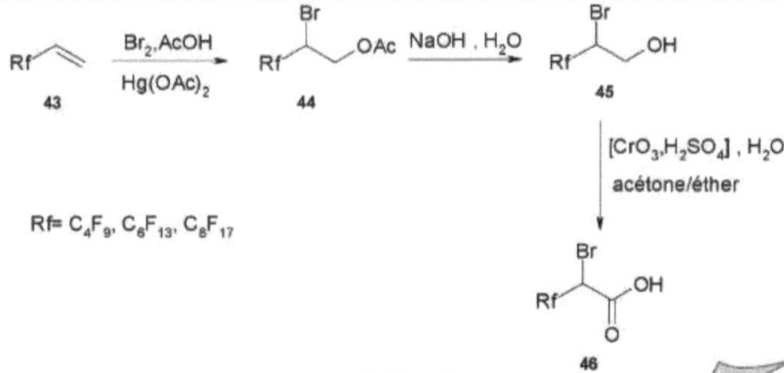

Schéma 19

II-4. Synthesis of a-bromoacids from Gem dicyano epoxide

Robert et al,[35] have described a new, simple and inexpensive synthesis method for obtaining a-bromoacids. This involves refluxing the gem dicyano epoxide of type <u>**47**</u> with bromohydric acid in tertrahydrofuran THF for three hours. Type <u>**49**</u> a-bromoacids were obtained selectively after hydrolysis in 47% chemical yield. The proposed method is easy to implement and the starting products are readily available **(Scheme 20)**.

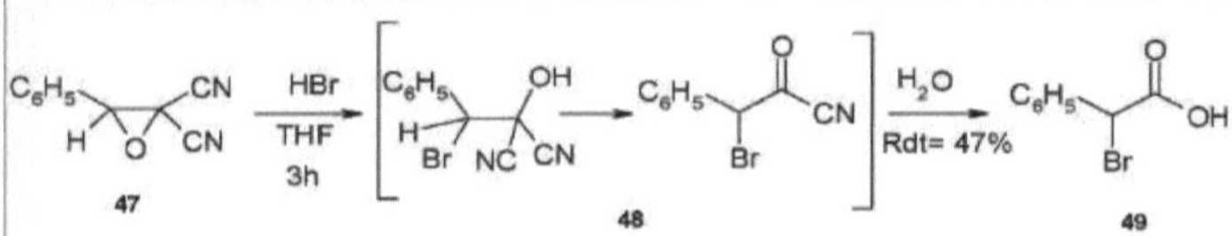

Schéma 20

II-5. Synthesis of a-bromo acids from amino acids

In recent years, the use of amino acids as starting materials for the preparation of chiral molecules has developed well[36] . In this context, Larcheveque et al,[37] have transformed (*L*)-serine <u>**50**</u> into the corresponding a-bromo acid **51**.

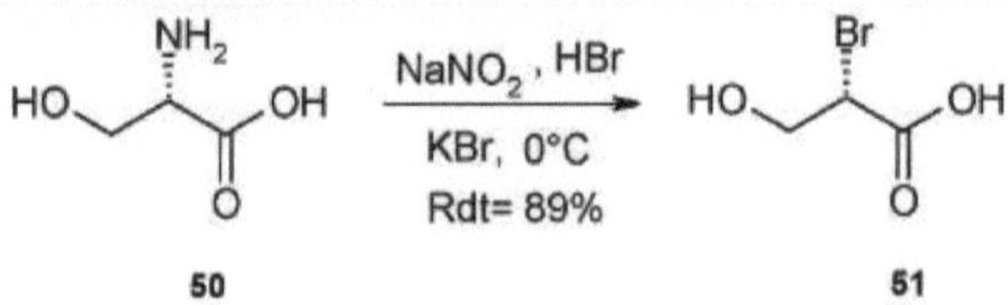

Schéma 21

A. Personal work

I. Synthesis of optically active a-bromoacids from natural amino acids and mechanistic study

In the course of this work, we prepared a series of six optically active a-brominated acids of the **53(a-f)** type in good chemical yields by nitrous deamination of naturally occurring α-amino acids at low temperatures. We applied this method to the following six amino acids:

(*L*)-alanine **52a**, (*L*)-valine **52b**, (*L*)-Leucine **52c**, (*L*)-isoleucine **52d**, (*L*)-phenylalanine **52e** and (*L, D*) Leucine **52f**. We used a saturated potassium bromide solution that allowed us to work at -13°C in order to maintain the stereospecificity of the reaction.

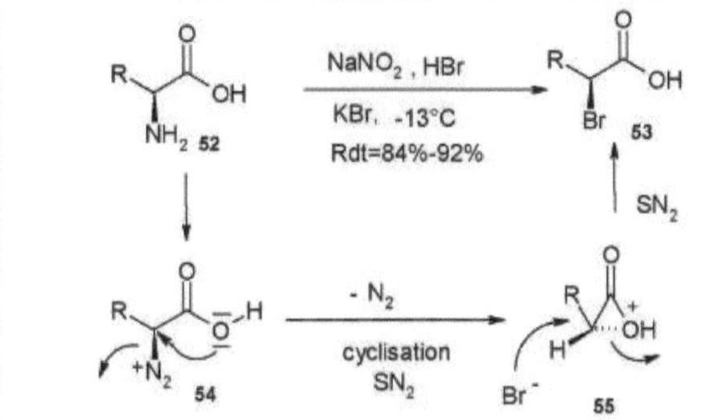

a : R=Me; **b** : R=i-Pr; **c** : R=i-Bu; **d** : R=Sec-Bu; **e** : R=Bn; **f*** : R=i-Bu

Schéma 22

The action of nitrous acid (HNO_2), generated *in situ* by the reaction of hydrobromic acid with sodium nitrite, leads to the formation of the corresponding diazonium salt **54**. This provides the unstable lactone **55**, which is formed after intramolecular cyclisation using an SN2 mechanism resulting in inversion of the configuration of the asymmetric carbon (*S R*). The opening of this a-lactone following SN2 nucleophilic substitution by bromide ions present in the reaction medium gives the correspondinga-bromoacids **53(a-f)** with a second inversion of the configuration of the asymmetric carbon which returns to its initial configuration (*R S*).

The chemical yields and physical aspects of the a-bromoacids prepared are given in the table below:

Table I

Chemical yields and physical aspects of prepared a-bromoacids 53(a-f)

Amino acids	Acids a-brominated	R	Yield (%)	Physical aspect
Alanine	**53a**	Me	70	Colourless oil
Valine	**53b**	i-Pr	98	Yellowish oil
leucine	**53c**	i-Bu	90	Yellowish oil
isoleucine	**53d**	Sec-Bu	91	Colourless oil
phenylalanine	**53e**	Bn	79	Yellowish oil
Leucine racemate	**53f**	i-Bu	93	Yellowish oil

1. Spectroscopic studies of a-bromoacids

The a-bromoacids prepared in this work were identified by H NMR^{1} and C NMR^{13} .

1.1-NMR spectroscopic properties H^{1}

The H NMR spectra^{1} of the prepared products are in agreement with the structures of the compounds prepared (see experimental section). Thus, analysis of the H NMR spectrum^{1} of compound **53d** recorded in deuterated chloroform ($CDCl_3$) at 300Hz, enabled us to detect the following signals:

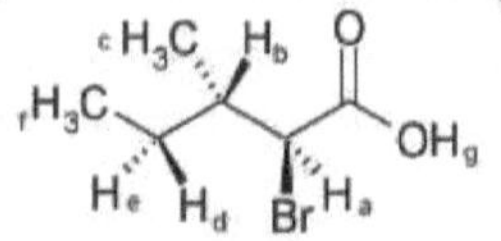

53d

(2S, 3S)-2-broimo-3-imethylpentano) / (l ue acid

- A triplet at 0.93 ppm 3H integration relating to the methyl protons (CH $)_{3f}$ with a coupling constant (*J*= 7.2 Hz)
- A doublet at 1.05 ppm 3H integration relating to the methyl protons (CH $)_{3c}$ with a coupling constant (*J*= 6.6 Hz).
- A multiplet between 1.26-1.38 ppm 1H integration relating to the H proton$_e$.
- A multiplet between 1.70-1.79 ppm 1H integration relating to the H proton$_d$.
- A multiplet between 2-2.09 ppm of 1H integration relating to the H proton$_b$.
- A doublet at 4.12 ppm 1H integration related to the H proton$_a$ with a coupling constant (*J*= 8.1 Hz).
- A broad singlet integrates a proton at 9.80 ppm corresponding to the proton of the acid function.

1. NMR spectroscopic properties C^{13}

Analysis of the^{13} C NMR spectrum recorded at 75MHZ with total proton decoupling of the same compound **53d** in deuterated chloroform revealed the following signals:

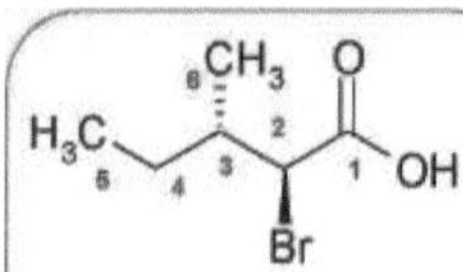

53d

- A signal at 10.10 ppm relating to carbon C_5 .
- A signal at 15.73 ppm relating to carbon C_6 .
- A signal at 25.71 ppm relating to carbon C_4 .
- A signal at 37.60 ppm relating to the asymmetric carbon C_3 .
- A signal at 51.95 ppm relating to carbon C_2 .
- A signal at 175.01 ppm relating to carbon C_1 of the carbonyl group (CO).

Chapter 2

Synthesis of optically active a'-brominated esters

A. General information and bibliographical references

The literature review showed that a-bromoesters are products of great synthetic interest and have been studied by several research groups.[38] In particular, a-bromoesters have been widely used as starting materials for many biologically active heterocyclic compounds.[39-49].

In this chapter, we describe some of the applications of a-bromoesters in heterocyclic synthesis, as well as their main synthetic routes, and then present our own work.

I. Synthetic interests

A wide range of products of biological interest have been prepared from a-bromoesters. Here are just a few examples:

I-1. Synthesis of benzyl cinnamates from a-bromoesters

Benzylcinnamates have made great progress in recent years thanks to their various biological activities: anti-inflammatory, anticancer[39a] and antibacterial[39b] In this context, a new method for synthesising benzylcinammate **58** was developed by Smita et al,[40] using the Wittig reaction. Treatment of aryl a-bromoester **56** with benzaldehyde **57** in the presence of triphenylphosphine in an aqueous sodium bicarbonate solution ($NaHCO_3$) leads to the formation of benzyl cinnamate **58** in good chemical yields **(Scheme 23).**

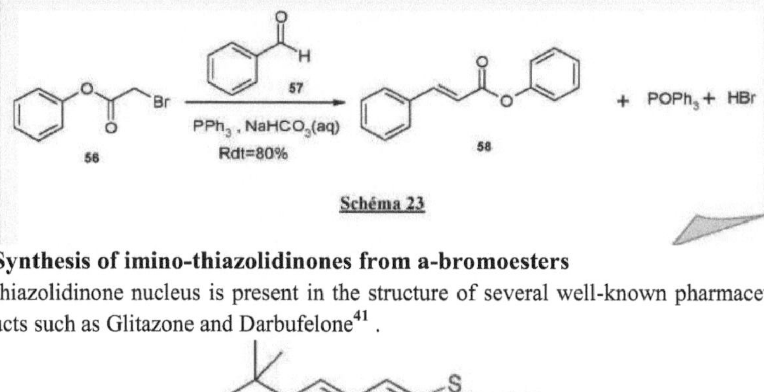

Schéma 23

I-2. Synthesis of imino-thiazolidinones from a-bromoesters

The thiazolidinone nucleus is present in the structure of several well-known pharmaceutical products such as Glitazone and Darbufelone[41].

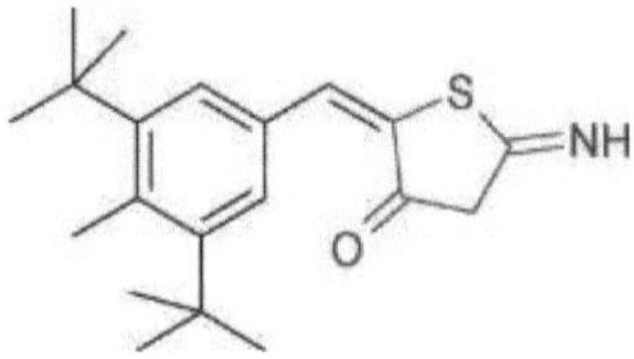

Darbufelone

Danis et al,[42] succeeded in preparing imino-thiazolidinones **61** by condensation of thiourea **59** with ethyl a-bromoacetate **60** in the presence of sodium acetate (NaOAc) in ethanol **(Scheme 24).**

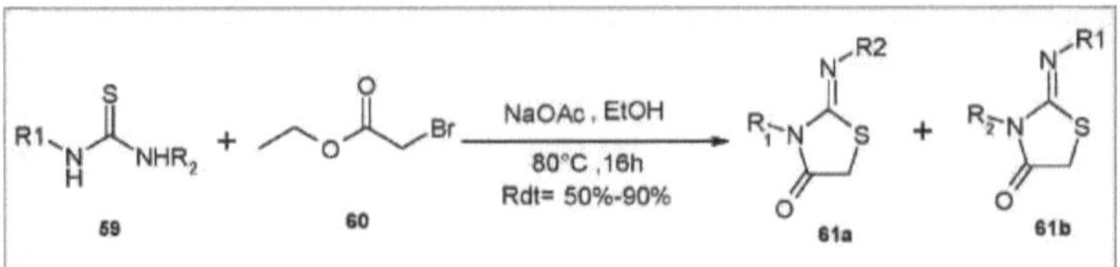

Schéma 24

I-3. Synthesis of 1,2,3-triazoles from a-bromoesters

There is a growing demand for the development of new antimicrobial agents due to the resistance developed by traditional antibiotics. For this reason, several syntheses have been made of 1,2,3-triazole derivatives with good antimicrobial activity.[43] Most recently, Hansen et al ([44]) prepared disubstituted 1,2,3-triazole derivatives **65** from ethyl bromoethanoate **62.** This reaction involves applying the 1,3-dipolar cycloaddition reaction of an organic azide **63** with alkynes **64**.

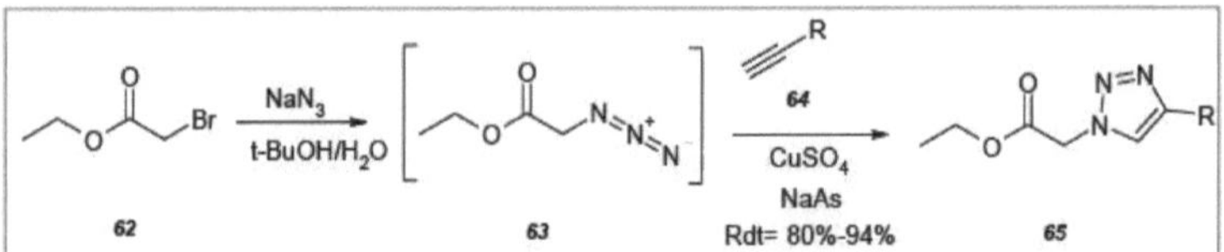

Schéma 25

I-4. Synthesis of quinoxalin-2-ones from a-bromoesters

Quinoxaline nuclei are present in the structure of many natural and synthetic compounds with interesting biological activities as insecticides, herbicides, antifungals[45] and antibacterials. The development of methods for synthesising these compounds has been the subject of several research projects. King et al,[46] described the synthesis of quinoxalin-2-one **70** by condensation of a-haloesters **68** with ortho-phenylenediamine **67** followed by oxidation with hydrogen peroxide **(Scheme 26).**

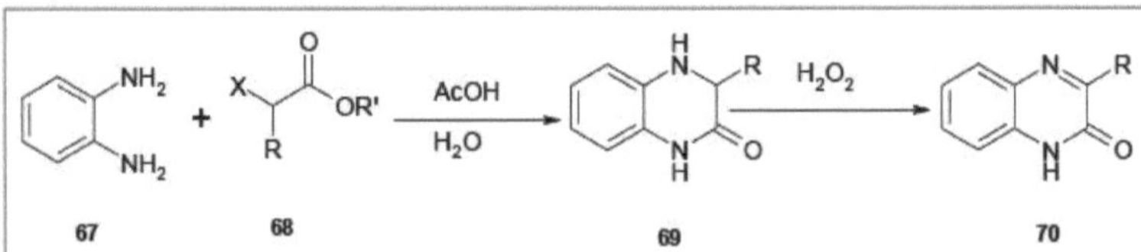

X=Cl, Br; R=H, alkyle; R'=alkyle

Schéma 26

I-5. Synthesis of *N-tosyl-1*,2,4-triazin-5-ones from a-bromoesters

Triazinone compounds have been shown to have various anticancer and anti-HIV pharmacological activities[47] . Recently, F. Allouche et al,[48] developed a method for the synthesis of *N-tosyl-1*,2,4-triazin-5-ones which consists of mole-by-mole condensation of

N-tosylamidrazones 71 with a-bromoesters 72 in THF in the presence of two moles of sodium hydride. The *N-tosyl-1*,2,4-triazin-5-ones 73 were obtained in average chemical yields of between 55% and 64% **(Scheme 27).**

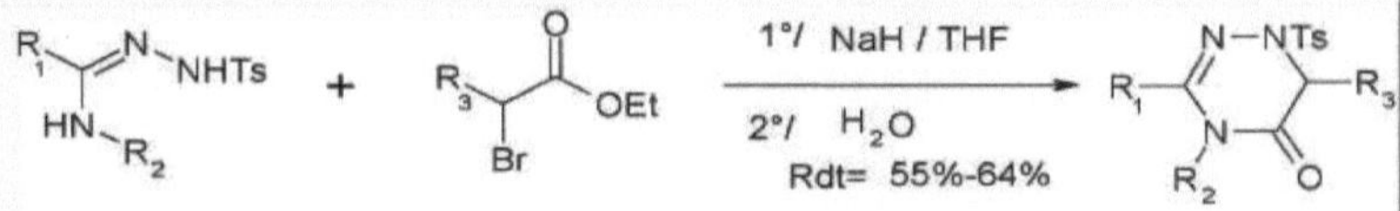

71 72 73

Schéma 27

I-6 Synthesis of imidazotriazol-5-ones from a-bromoesters

Work carried out by F. Allouch et al.[49] , led to the synthesis of new nitrogen-containing biheterocyclic compounds of type **76**. Treatment of 5-amino-1-phenyl-1,2,4- triazoles 74 with a-bromoesters 75 in the presence of two equivalents of sodium hydride in THF gave imidazotriazol-5-ones 76, which exhibit antimutagenic activity.

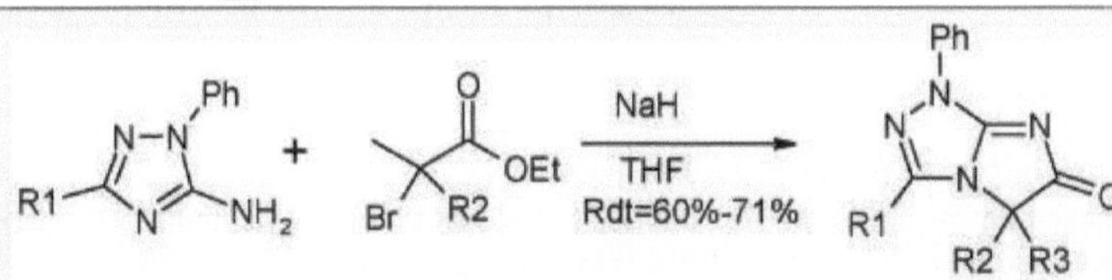

74 75 76

Schéma 28

The following summary diagram demonstrates the synthetic importance of a-bromoesters in providing access to numerous biologically active heterocyclic compounds.

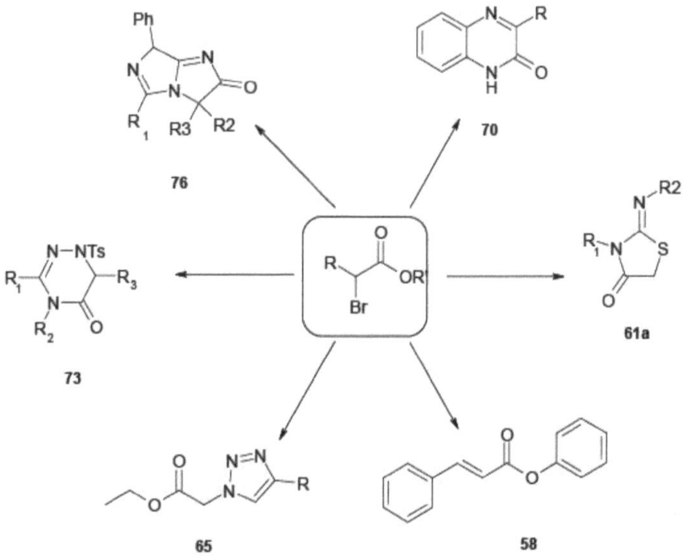

II. Main routes for preparing a-bromoesters

The literature review has shown that several methods for synthesising a-bromoesters have been developed, some of which are listed below:

II-1. Synthesis of a-bromoesters by ester bromination

II-1.1. Bromination by N-bromosuccinimide

Methyl 2-bromo-2-phenoxyacetate **78** was obtained in 87% chemical yield by bromination of methyl phenoxyacetate **77** with N-bromosuccinimide to benzene reflux[50] .

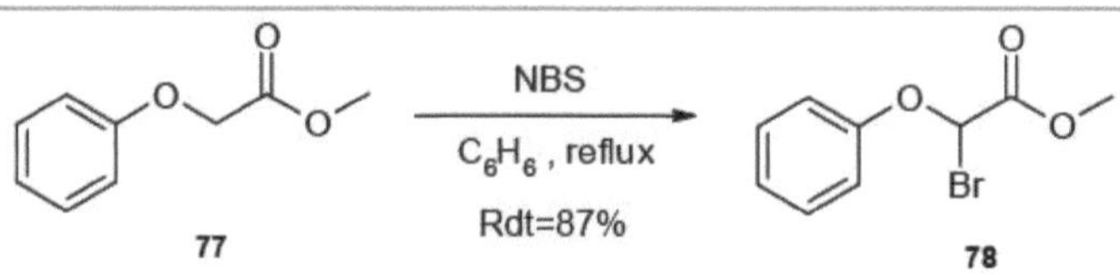

Schéma 29

II-1.2. Bromination with tetrabromomethane

Treatment of ester **79** with lithium diisopropylamide (LDA) at -78°C yields the corresponding carbanion **80**, which reacts rapidly with one equivalent of tetrabromomethane to form the a-bromoesters **81** in good chemical yields[51] **(Scheme 30).**

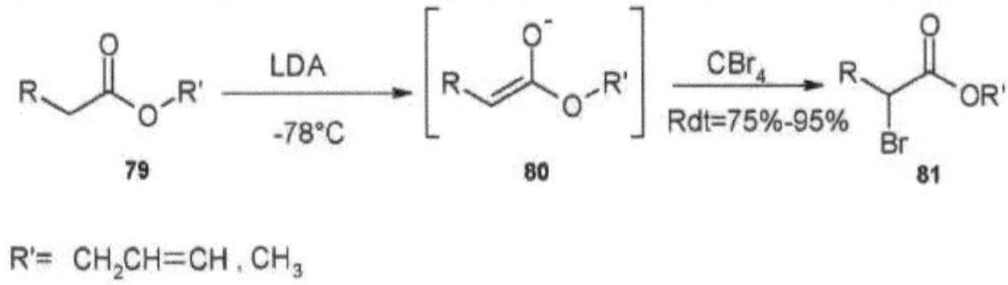

Schéma 30

II.2. . Synthesis of a-bromoesters from aryldiazonuim salts

Minrou et al,[52] repeated Meerwein's reaction[53] to prepare α-bromo esters **84**. The reaction of arylamines **82** with sodium nitrite and bromohydric acid leads to the aryldiazonuim salt, which reacts with butyl acrylate **83** in the presence of a catalytic amount of copper bromide to form the corresponding a-bromoester **84** **(Scheme 31).**

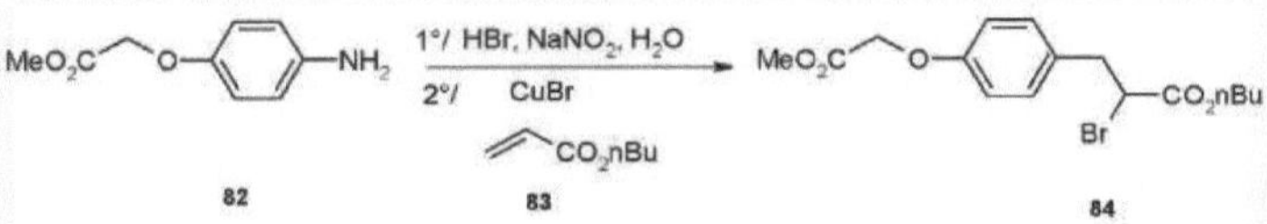

Schéma 31

II-3. Synthesis of a-bromoesters from β-ketoesters

T. Aoyama et al,[54] described a new method for the preparation of a-bromoesters. This method involves reacting β-ketoesters **85** with copper dibromide and sodium carbonate (supported on alumina) under reflux of benzene for one hour to give a-bromoesters **86**. This reaction has the advantage of being rapid and efficient. **(Scheme 32).**

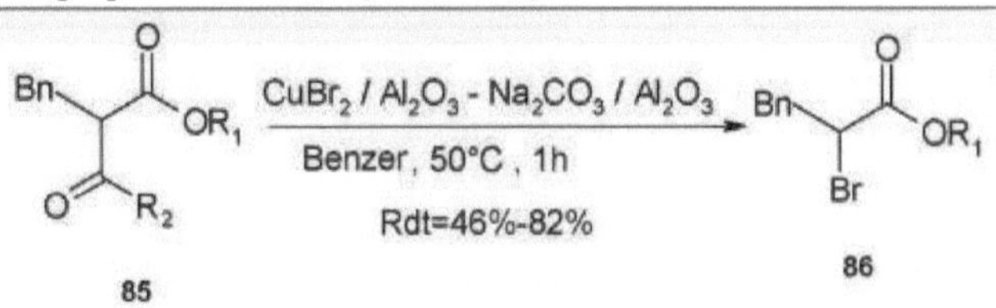

Schéma 32

II-4 Synthesis of α-bromoesters from carboxylic acids

II-4.1. Passage through acid chloride

Katrina et al,[55] used the method of Schwenk and Papa[56] to prepare a-bromoester **88**. Treatment of carboxylic acid **87** with thionyl chloride at a temperature of 70°C yields acid chloride, which reacts with bromine to form brominated acid chloride. The latter reacts readily with ethanol to form the a-bromoester **88** in 81% yield **(Scheme 33).**

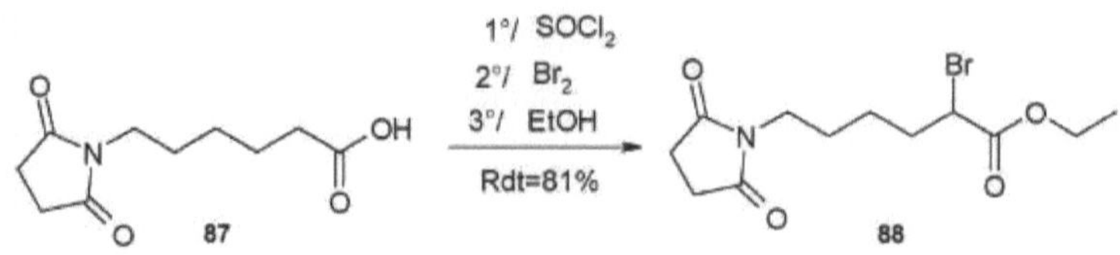

Schéma 33

II-4-2. Passage through acid bromide

Ogata et al,[57] have described another synthetic route which gives a-bromoesters **91** in chemical yields of between 78% and 95%. This involves reacting carboxylic acids **89** with dibromine in the presence of a catalytic amount of chlorosulphonic acid to give the brominated acid **90**, which is refluxed in methanol for two hours to give the a-bromoester **91** **(Scheme 34)**.

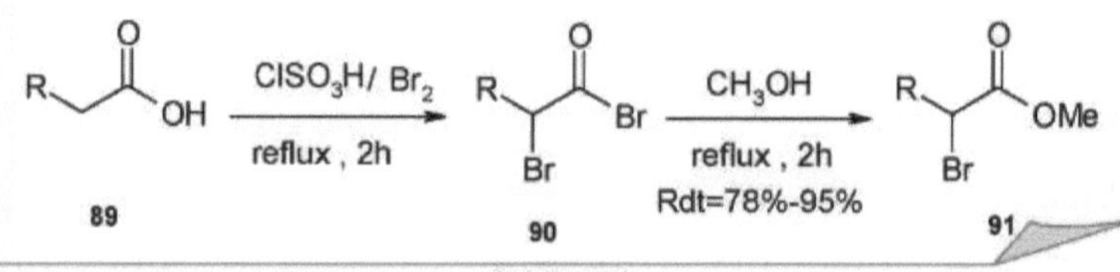

Schéma 34

B. Personal work

I. Synthesis of optically active α-bromoesters and mechanistic study

The most widely described method in the literature for converting carboxylic acids to the corresponding esters is via acid chloride. In this work, we adopted this method to prepare a series of six optically active a-bromoesters from the corresponding α-bromo acids prepared in the first chapter.

Treatment of the α-bromoacids **53(a-f)** with thionyl chloride used as reagent and solvent afforded the corresponding acid chlorides after 4 hours heating at 50°. These were then converted to the corresponding **93(a-f)** esters by treatment with methanol at 0°C in the presence of one equivalent of triethylamine. This method produces the a-bromoesters **53(a-f)** in good chemical and optical yields. The reaction sequence is summarised in the following diagram:

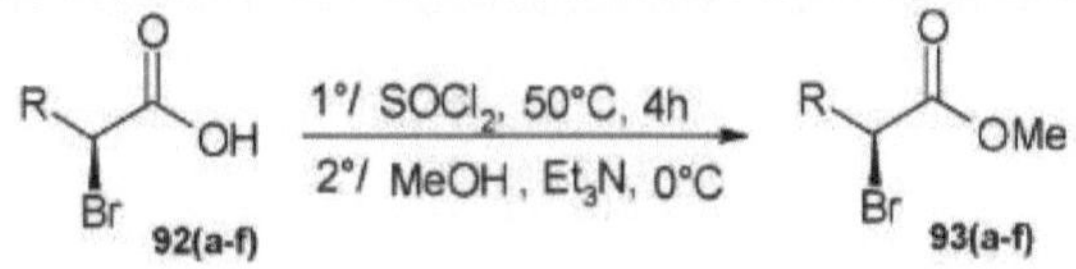

a : R=Me; **b** : R=i-Pr; **c** : R=i-Bu; **d** : R=Sec-Bu; **e** : R=Bn; **f*** : R=i-Bu

<u>**Schéma 35**</u>

The mechanism of this reaction is as follows:

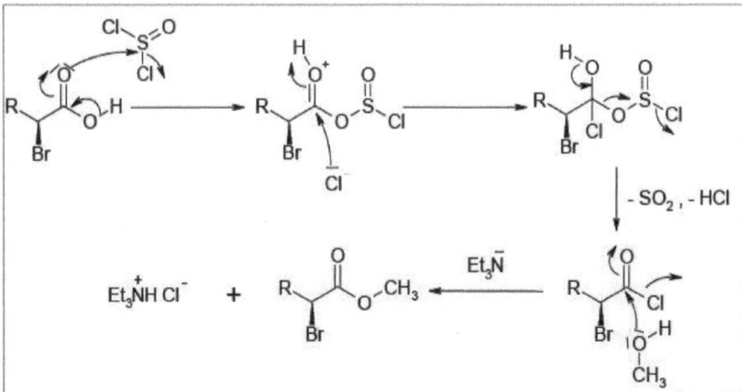

The chemical yields and physical aspects of the products prepared are shown in the table below:

Table II

<u>*Chemical yields and physical aspects of a-bromoesters prepared 93(a-f)**</u>

93R	Yield (%)	Appearance	
93a	Me	78	Colourless oil
93b	i-Pr	85	Yellowish oil
93c	i-Bu	83	Yellowish oil
93d	Sec-Bu	90	Colourless oil
93e	Bn	76	Yellowish oil
93f*	i-Bu	80	Yellowish oil

11. Spectroscopic properties of prepared a-bromoesters

The a-bromoesters were identified by NMR spectroscopy at ^{1}H and ^{13}C.

II-1. <u>NMR spectroscopic properties H^{1}</u>

NMR spectroscopy at ^{1}H confirmed the structures of the a-bromoesters prepared (see experimental section). As an example, let's look at the case of methyl (*S*)-2-bromo-4-methylpentanoate <u>**93c**</u> :

* prepared from (*D*, *L*) Leucine.

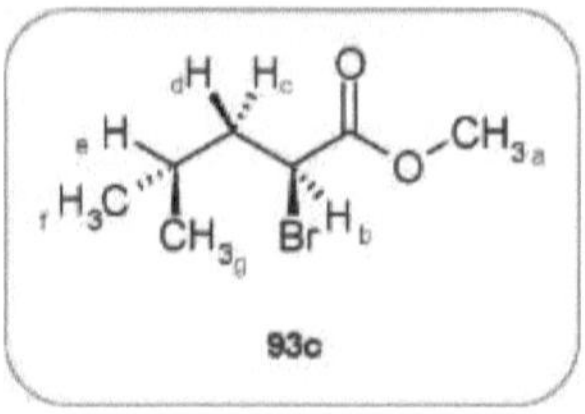

Methyl (2S)-2-bromo-4-methylpentanoate

The NMR spectrum[1] H of compound **93c (Figure 5)** recorded at 300 MHz in deuterated chloroform shows the following signals:

- A doublet at 0.89 ppm 3H integration related to the methyl protons $(CH\ )_{3f}$ (J= 6.6MHz).
- A doublet at 0.93 ppm 3H integration related to the methyl protons $(CH\ )_{3g}$ (J= 6.6MHz).
- A multiplet between 1.73-1.80 ppm 1H integration relating to the H_e proton.
- A multiplet between 1.88-1.93 ppm 2H integration relating to the H protons_c and H_d .
- A singlet at 3.77 ppm 3H integration relating to the protons of the methoxy group (OCH_3).
- A triplet at 4.28 ppm integrating a proton relative to the H proton_b .
- **I.2 NMR spectroscopic properties C[13]**

The total proton decoupling NMR spectrum[13] C of the same compound **93c (Figure 6)** recorded in deuterated chloroform at 75 MHz confirms the structure based on the following signals:

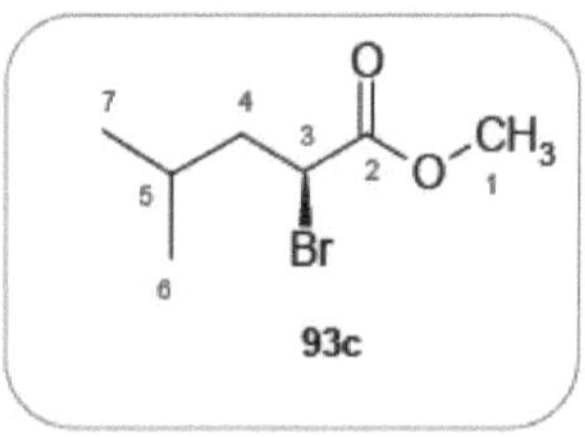

- A signal at 21.03 ppm relating to carbon C_7 .
- A signal at 21.82 ppm relating to carbon C_6 .
- A 25.83 ppm carbon signal C_5 .
- A 43.03 ppm carbon signal C_4 .
- A 43.82 ppm carbon signal C_3 .
- A 52.37 ppm carbon signal C_1 .
- A signal at 170.06 ppm relating to carbon C_2 .

Chapter 3

Synthesis of new chiral 1,2,4-oxadiazin-5-ones

A. General information and bibliography

Nitrogen heterocycles are present in a number of natural compounds that play an important role in cellular metabolism. Oxadiazinones are one of several classes of these compounds known for their biological activities[58] and can be found as central nervous system stimulants[59] , oxidase inhibitorsB (MAO-B)[60] and antibacterial[58-61] .

On the other hand, oxadiazinone derivatives are excellent intermediates used in symmetrical and asymmetrical organic synthesis[62] . These six-membered heterocycles present several varieties depending on the relative positions of the oxygen atom and the two nitrogen atoms[63] . Examples include: 1,2,4-oxadiazin-3-ones **94**[63] ; 1,2,4-oxadiazin-5-ones[64] **95**; 1,2,4-oxadiazin-6-ones[64] **96** and 1,3,4-oxadiazin-2-ones[65] **97**. The interest aroused by this class of compounds in various fields led us to direct our work towards the synthesis of new chiral 1,2,4-oxadiazin-5-ones.

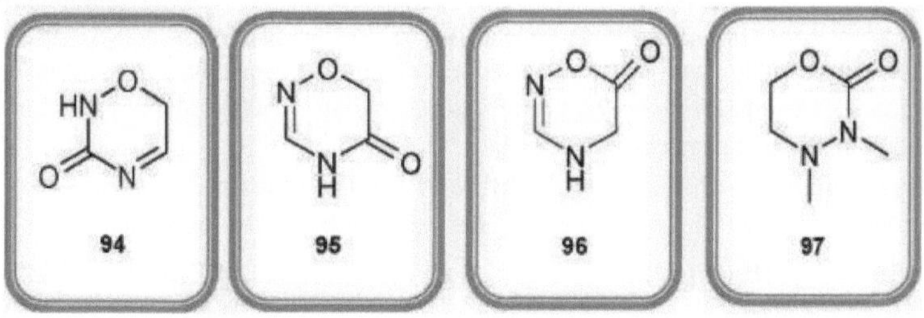

In what follows, we present a few significant examples of oxadiazinone applications and their preparation methods. We will then present our method, which consists of condensing the a-bromoesters previously prepared with (2 *5*)-1-benzensulphonylpyrrolidin-2-carboxime derived from proline.

I. Synthetic benefits of oxadiazinones

Optically active oxadiazinones are widely used in various areas of organic synthesis. Some examples are given below:

I-1. Diastereoselective aldolisation reaction

The 1,3,4-oxadiazin-2-ones **98** have been successfully used as chiral auxiliaries in diastereoselective aldolisation reactions[66] following the method described by Hitchcock et al,[67] **(Scheme 36).**

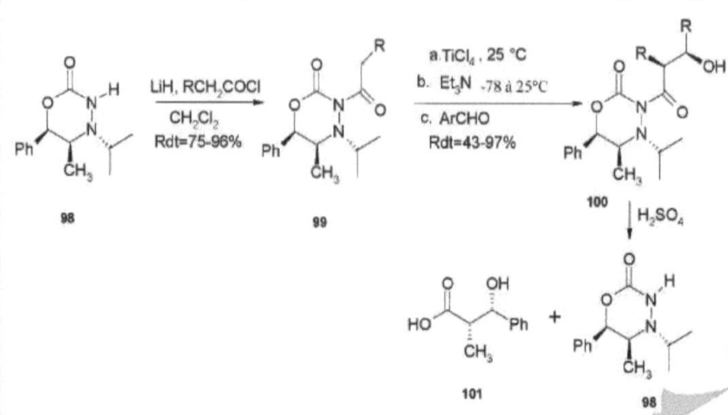

Schéma 36

This diastereoselective aldolisation is carried out in three stages. Type **98** 3,4,5,6-tetrahydro-1,3,4-oxadiazin-2-ones are acylated with propionyl chloride in a basic medium to give *N-3-propionyl-3*,4,5,6-tetrahydro-1,3,4-oxadiazin-2-one **99** which is treated with titanium tetrachloride to give chlorotitanium enolate. The latter reacts with trimethylamine and an aromatic aldehyde to form the intermediate of type **100** with a majority (*2S* ,*3S*) configuration which is subsequently hydrolysed in an acid medium to regenerate the starting 3,4,5,6-tetrahydro-1,3,4-oxadiazin-2-one **98**, as well as 3-Hydroxy-2-methyl-3-phenylpropionic acid **101** with good diastereoisomeric excesses.

I-2. Asymmetric 1,3-dipolar cycloaddition of 1,3,4-oxadiazin-2-ones

Husson et al,[68] used 1,3,4-oxadiazin-2-ones as synthons for asymmetric 1,3-dipolar cycloaddition reactions. Condensation of 1,3,4-oxadiazin-2-ones **102** with dimethylacetalbenzaldehyde in the presence of paratoluene sulphonyl acid APTS leads to the azomethine imine intermediate of type **103**. The latter reacts directly with diethylacetylenedicarboxylate to form compounds **104** as a single diastereoisomer in 35% yield **(Scheme 37)**.

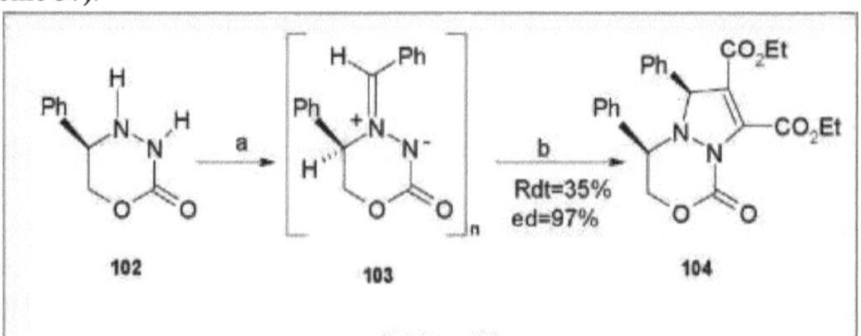

Schéma 37

Reagents and conditions: **a)** $PhCH(OMe)_2$, APTS, toluene, 70°C; **b)** EtO C-CH=CH-CO_{22} Et.

I-3. Preparation of new bicyclic hydrazinolactams

Husson et al,[69] showed that the condensation of 1,3,4-oxadiazin-2-ones of type **105** with bromobutyl chloride in the presence of potassium carbonate led to the formation of new bicyclic hydrazinolactams of type **106** with a chemical yield of 81% **(Scheme 38)**.

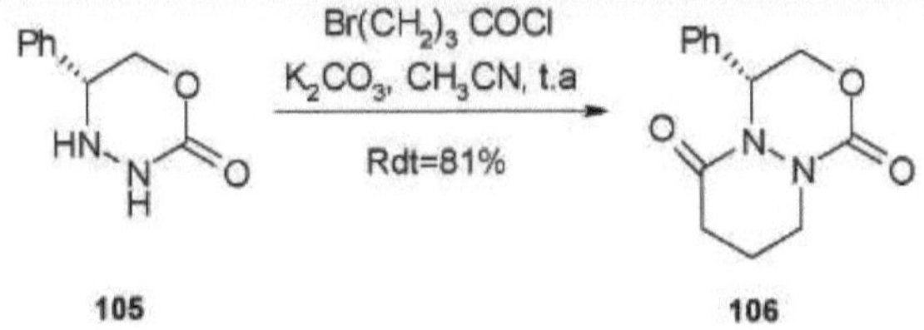

Schéma 38

II. Main routes for preparing oxadiazinones

The presence of the oxadiazinone ring in the skeleton of many molecules with confirmed biological activity in various fields has encouraged researchers to develop new methods for synthesising this type of compound.

II-1. Synthesis of oxadiazinones from ephidrine

J. R. Burgeson et al,[66] described the preparation of 1,2,3-oxadiazin-2-ones of type **110** by treating ephidrin **107** with acetone followed by sodium borohydride to form intermediate **108** in 72% yield. The reaction of the latter with

Sodium nitrite and hydrochloric acid give compound **109.** This compound is reduced by the action of lithium hydride and then cyclised in the presence of diethylcarbonate and lithium hydride to give the oxadiazinones **110.**

(Diagram 39).

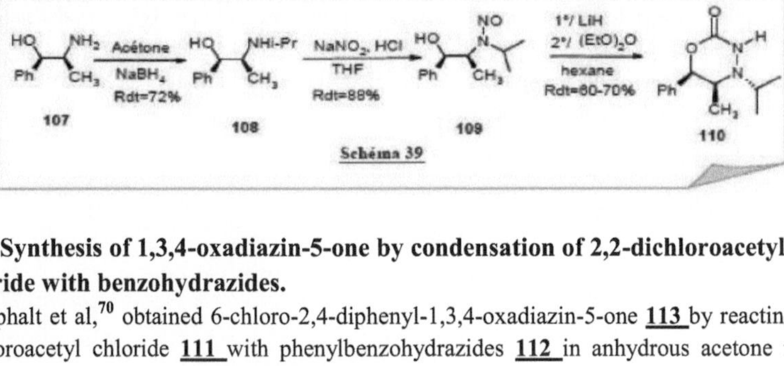

Schéma 39

II-2. Synthesis of 1,3,4-oxadiazin-5-one by condensation of 2,2-dichloroacetyl chloride with benzohydrazides.

Westphalt et al,[70] obtained 6-chloro-2,4-diphenyl-1,3,4-oxadiazin-5-one **113** by reacting 2,2-dichloroacetyl chloride **111** with phenylbenzohydrazides **112** in anhydrous acetone in the presence of potassium carbonate **(Scheme 40)**.

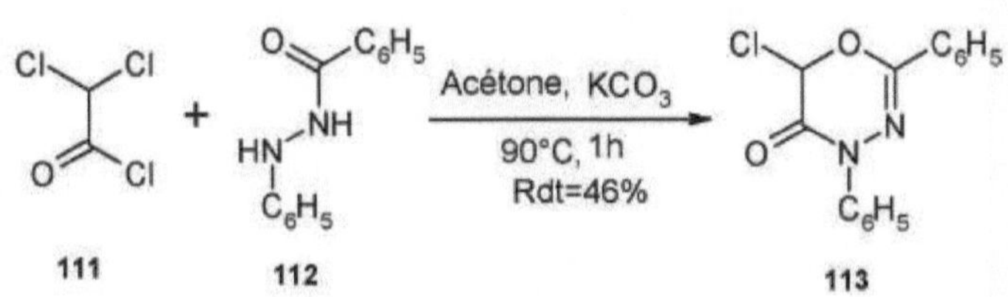

Schéma 40

II-3. Synthesis of substituted 1,3,4-oxadiazin-5-ones

S. Y. Ke et al,[58] have been interested for several years in the preparation of 1,3,4-oxadiazin-5-ones and the study of their biological activities. Treatment of the various benzoates **114** with hydrazine hydrate in ethanol leads to the formation of hydrazides **115**. Condensation of the latter with acrylonitrile or n-butyl acrylate in ethanol at 60°C results in intermediates **116** which undergo cyclisation with chloroacetyl chloride in the presence of potassium carbonate to give 1,3,4-oxadiazin-5-ones of type **117**.

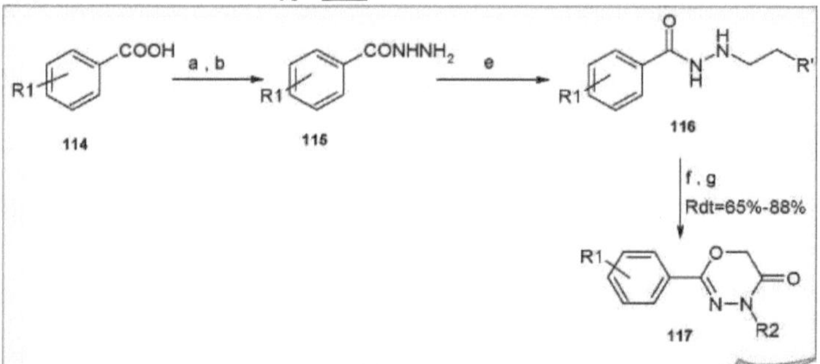

Schéma 41

Reagents and conditions: **a)** EtOH, H_2SO_4 ; **b)** 5 equiv $NH_2NH_2 . H_2O$, reflux for 8 h; **e)** EtOH, acrylonitrile or n-butyl acrylate, at 60°C for 30 to 48 h; **f)** 1.2 equiv $ClCH_2COCl$, $CHCl_3$, reflux for 1 h; **g)** 5 equiv K_2CO_3, EtOH, reflux for 0.5 to 2 h.

II-4. Synthesis of 2,5-diaryl-1,3,4-oxadiazin-6-ones

Tîntas et al,[71] succeeded in preparing new oxadiazinones of type **121**. Condensation of phenylglyoxylic acid **118** with aroylhydrazines **119** leads to aroylhydrazone **120**. This undergoes esterification in the presence of DCC, followed by intramolecular cyclisation to give the desired oxadiazinones **(Scheme 42)**.

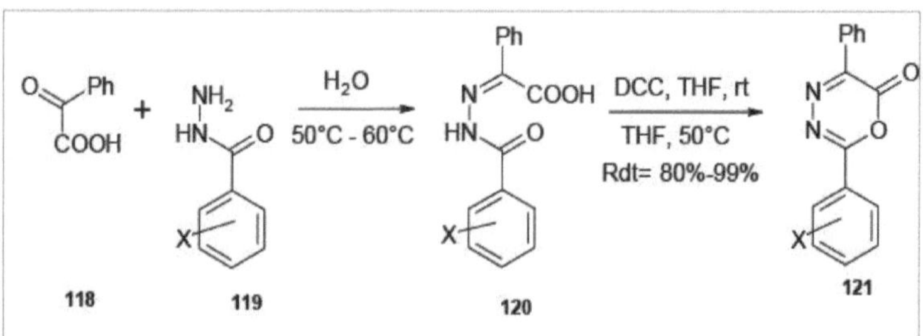

a) X=m-Br; b) X=m-Cl; c) X=m-OMe; d) X=p-tBu.

Schéma 42

II-5. Synthesis of 1,2,4-oxadiazin-6-ones by condensation of aromatic nitrile oxides with a-amino esters

The first synthesis of 1,2,4-oxadiazin-6-ones was carried out by Hussein et al,[63] in 1984. This method consists of condensing aromatic nitrile oxides **122** with a-amino esters **123** leading to intermediates **124** which cyclize to give 1,2,4-oxadiazin-6-ones **125** in yields of between 12% and 73% **(Scheme 43)**.

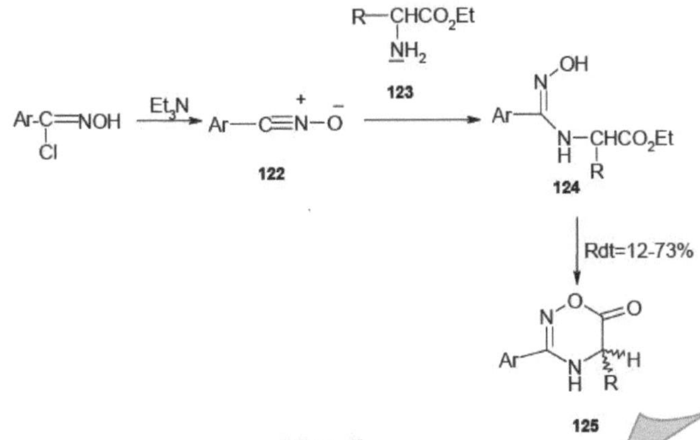

Schéma 43

II-6. Synthesis of 1,2,4-oxadiazin-6-one from oxadiazoles

A new method for preparing 1,2,4-oxadiazin-6-ones was developed by Piccionello et al,[72] . The action of hydroxylamine on *oxadiazole5* **126** in the presence of a strong base gives intermediates **127** which rearrange to give 1,2,4-oxadiazin-6-ones **128** as shown in the following diagram.

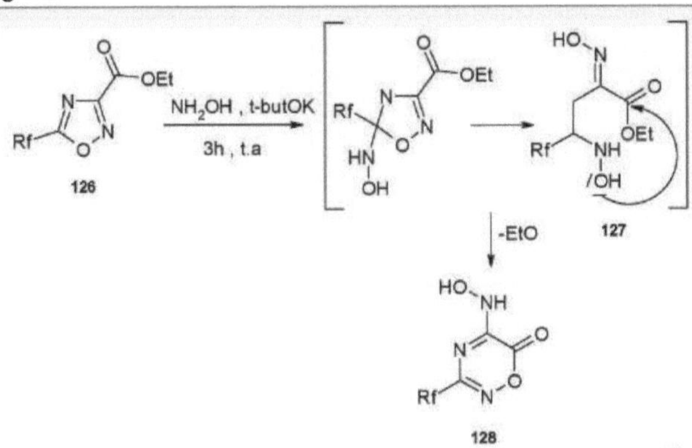

Schéma 44

B. Personal work

I. Synthesis of 1,2,4-oxodiazin-5-ones and mechanistic study

Amidoximes are compounds widely used in heterocyclic synthesis[73] .
have two nucleophilic centres capable of reacting with bi-electrophilic reagents.
In the course of this work, we prepared a new series of 1,2,4-oxodiazin-5-ones by condensation of 1-benzenesulfonylpyrrolidin-2-carboxamidoximes **129** prepared from proline[B. * * * * 13] with α-bromoesters **93(a-f)** in the presence of two equivalents of sodium hydride **(Scheme 46)**. All 1,2,4-oxodiazin-5-ones are obtained as a mixture of two diastereoisomers.

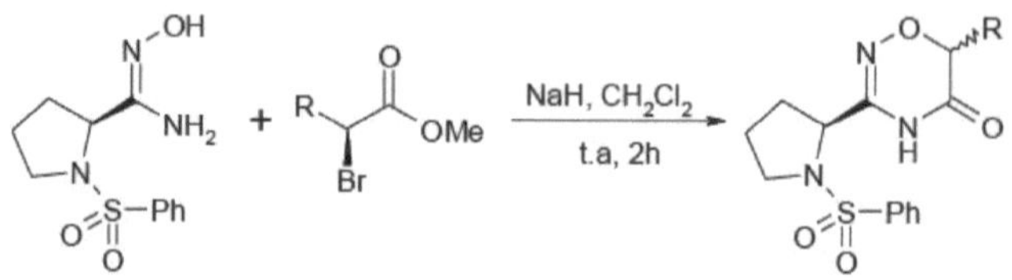

129 **93(a-e)** **130(a-e)**

a : R=Me ; **b** : R= i-Pr ; **c** : R= i-Bu ; **d** : R= Sec-Bu ; **e** : R= Bn.

Schéma 46

Reaction mechanism

The α-bromo ester has two neighbouring electrophilic centres which are slightly different from the electrophilic point of view and which react by two possible mechanisms with amidoxime **129** to give the chiral 3-(1-phenylsulphonyl-2-pyrrolidyl)-1,2,4-(4H)-oxadiazin-5-ones **130(a-e)**

The first mechanism involves nucleophilic attack of the free oxygen doublet of 1-benzenesulfonylpyrrolidin-2-carboxamidoxime **129** on the bromine-bearing tertiary carbon to form intermediate **131**. The latter undergoes intramolecular cyclisation in the presence of sodium hydride following nucleophilic attack of the free nitrogen doublet on the carbonyl of the ester function and the departure of the methoxy group **(Scheme 47)**.

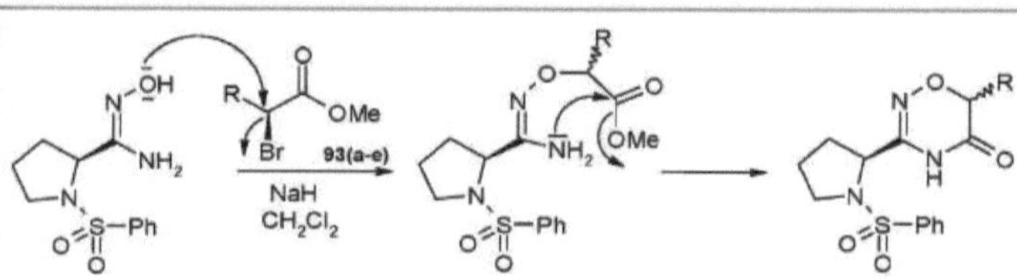

129 **131(a-e)** **130(a-e)**

Schéma 47

The second mechanism begins with a nucleophilic attack of the oxygen doublet of 1-benzenesulfonylpyrrolidin-2-carboxamidoxime **129** on the carbonyl of the ester function and the departure of the methoxy group and consequently the formation of intermediate **132**. The latter undergoes a rearrangement by attack of the free nitrogen doublet on the carbonyl carbon and subsequently the formation of intermediate **133**. At this stage, the desired product **130** is obtained by nucleophilic attack of oxygen on the bromine-carrying carbon and the departure of the bromide ion (Br^-) **(Scheme 48)**.

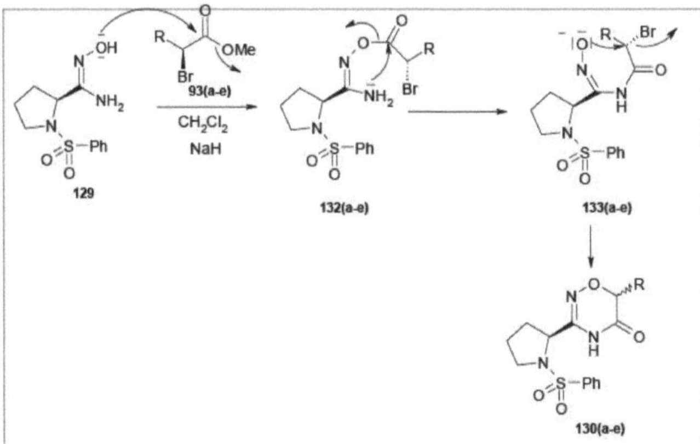

<u>Schéma 48</u>

The chemical yields of 3-(1-phenylsulfonyl-2-pyrrolidyl)-1,2,4-(4H)-
The **<u>130</u>(a-e)** 6-substituted oxadiazin-5-ones obtained are shown in the following table:

Table III

<u>Chemical yields and physical aspects</u>
<u>6-substituted 3-(1-phenylsulfonyl-2-pyrrolidyl)-1,2,4-(4H)-oxadiazin-5-ones **130(a-e)**</u>

<u>130</u>	R	Yields (%)	Aspect
130a	Me	78	Solid white
130b	i-Pr	80	Solid white
130c	i-Bu	83	Solid white
130d	Sec-Bu	89	Solid white
<u>130^{e}</u>	Bn	75	Solid white

Recrystallisation solvent (chloroform-petroleum ether).

II. Spectroscopic studies

The 3-(1-phenylsulfonyl-2-pyrrolidyl)-1,2,4-(4H)-oxadiazin-5-ones **<u>130</u>(a-e)** were identified by NMR spectroscopy at^{1} H and^{13} C.

1. <u>NMR spectroscopic properties H^{1}</u>

The NMR spectra at^{1} H confirmed the structures of the 1,2,4-oxadiazin-5-ones **<u>130</u>(a-e)** prepared (see experimental section). As an example, let's look at the case of compound **<u>130b</u>**, which was isolated in the form of a mixture of two diastereoisomers, as demonstrated by the duplication of certain signals.

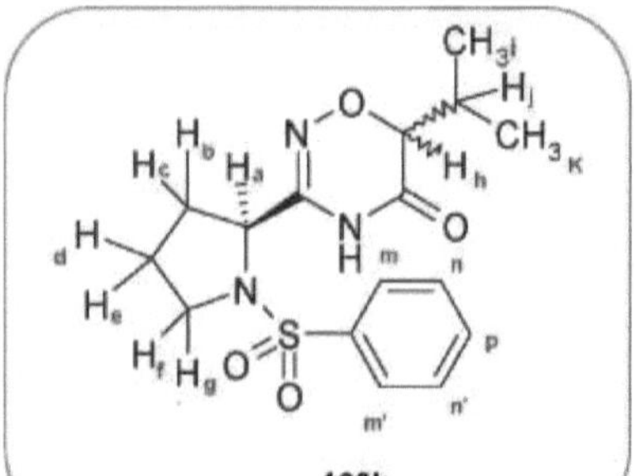

130b

(2'S, 5R) and (2'S, 5S) 6-isopropyl-3-(1'-benzenesulfonylpyrrolidin-2'-yl)-4,6-dihydro-1,2,4-oxadiazin-5-one: 130b

The NMR spectrum[1] H of compound **130b** recorded at 300Hz in the $CDCl_3$ shows the following signals:

- A 6H integration multiplet between 1.06 and 1.14 ppm corresponding to the methyl protons $(CH_3)_k$ and $(CH_3)_j$,
- A 2H integration multiplet between 1.64 and 1.71 ppm corresponding to the He and Hd protons.
- A 1H integration multiplet at 1.95 ppm corresponding to the H proton$_i$.
- A 2H integration multiplet between 2.30 and 2.39 ppm corresponding to the Hf and Hg protons.
- A 1H integration multiplet at 3.26 ppm corresponding to the H_c or H_b proton.
- A 1H integration multiplet at 3.59 ppm corresponding to the H_b or H_c proton.
- a 1H integration doublet at 3.95 ppm corresponding to the Hh proton.
- A split doublet of 1H integration at 4.28 ppm corresponding to the H proton$_a$ *(J=7.5 Hz; J=3.6 Hz)*.
- A 2H integration triplet at 7.60 ppm corresponding to the H protons$_n$ and H_n', (*J*=7.8 Hz).
- A 1H integration doublet at 7.69 ppm corresponding to the H proton$_p$, (*J*=7.5 Hz).
- A 1H integration doublet at 7.89 ppm corresponding to the H protons$_m$ and H_m', (*J*=8.7Hz).
- A broad 1H integration band at 8.83 ppm corresponding to the NH

2. NMR Spectroscopic Properties C^{13}

Analysis of the[13] C NMR spectrum of the same compound, recorded at 75 MHz in CDC13 with total proton decoupling, revealed the following signals:

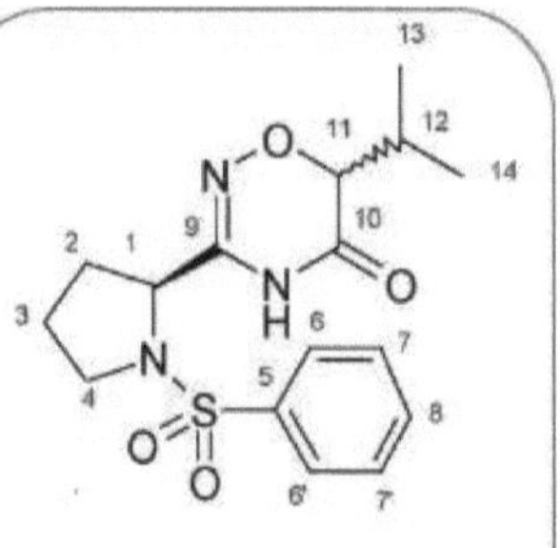

130b

- A signal at 16.72 ppm corresponding to carbon C_{13} .
- A signal at18.25 ppm corresponding to carbon C_{14} .
- A signal at 23.96 ppmcorresponding to the C carbon$_3$.
- A signal at 26.96 ppmcorresponding to the C carbon$_{12}$.
- A signal at 28.95 ppmcorresponding to the C carbon$_4$.
- A signal at 49.25 ppmcorresponding to the C carbon$_2$.
- A signal at 58.11ppm corresponding to carbon C_1 .
- A signal at 80.36 ppm corresponding to carbon C_{11} .
- Signals at 127.01; 127.33; 128.83; 129.03; 133.20; 135.04 ppm corresponding respectively to the aromatic carbons $C_{6,6'}$; $C7_{,7'}$; C8; C_5.
- A signal at 150.18 ppm corresponding to carbon C_9 .
- A signal at 165.81 ppm corresponding to carbon C_{10} .

References

[1] (a) Ager, D. J.; Mole, S. J. *Tetrahedron Lett*, **1988**, 29, 4807-4810.
(b) Gros, C.; Bonni, G. *L'actualité Chimique*, **1995**, 3, 9-15.

[2] Jackson, B. G.; Gardner, I. P.; Heath, P. C. *Tetrahedron Lett*, **1990**, 31, 6317-6320.

[3] Clayden, J.; Greeves, N.; Warren, S.; Wothers, P. "*General reviews on asymmetric synthesis",* **2002**, 45, 1219.

[4] Kagan, H.: "*Asymmetric catalysis", Pour la science*, Paris, **1992**, 172, 42.

[5] Haeseler, G.; Tetzlaff, D.; Dengler Bufler, J.; R.; Münte, S.; Hecker, H.; Leuwer, M. *Anesth. Analg*, **2003**, 96, 1019.

[6] (a) Ghanem, A.; Aboul-Enein, H. *Chirality*, **2005**, 17, 1-15.
(b) Ebbers, E. J.; Ariaans, G. J. A.; Houbiers, J. P. M.; Bruggink, A.; Zwanenburg, B. *Tetrahedron*, **1997**, 53, 9417-9476.

[7](a) Stork, G.; Nakahara, Y.; Greenlee, W. J. *J. Am. Chem. Soc.* **1978**, 100, 7775-7777.
(b) Overhand, M.; Hecht, S. M. *J. Org Chem*, **1994**, 194, 59, 4721-4722.

[8] Swinbourne, J. F.; Hunt, H. J.; Klinkert, G. *Adv. Heterocycl. Chem*, **1979**, 23, 103-170.

[9]Aliyenne, A. O.; Kharia, J. E.; Kraïm, J.; Kacem,Y.; Ben Hassine, B. *Tetrahedron Lett,* **2008**, 49,1473-1475.

[10]Aliyenne, A, O; Kharia, J.E; Kraïm, J.; Kacem, Y.; Ben Hassine, B. *Tetrahedron Lett*, **2006**, 47, 6405-6408.

[11] Kacem, Y.; Ben Hassine, B. *Tetrahedron Lett*, **2013**, 54, 4023-4025.

[12]Tka, N.; Ben Hassine, B.; *Synth. Commun*, **2010**,40, 3168-3176.

[13]Tka, N.; Ben Hassine, B. *Synthetic Commun.* **2012**, 42, 828-835.

[14] Park, Y. S. *Tetrahedron: Asymmetry*, 2009, 20, 2421-2427.

[15] Boschi, F.; Camps, P.; Franchini, M. C.; Torrero, D. M.; Ricci , A.; Sánchez, L. *Tetrahedron: Asymmetry*, **2005**, 16, 3739-3745

[16] Fairlie, D. P.; Abbenante, G.; March, D. R. *Curr. Med. Chem*, **1995,** 2, 654-686.

[17] Fischer, E.; Mechel, L. V. *Chem. Ber.* **1916**, 49, 1355-1366.

[18] Izumiya, N.; Nagamatsu, A. *Bull. Chem. Soc.* **1952**, 25, 265-267.

[19] Tka, N.; Kraiem, J.; Ben hassine, H. *Synth.Commun.* **2013**, 43, 735-743.

[20] (a) Camri, A. ; Pollak, G. *J.Org. Chem*, **1960**, 25,44-46.
(b) Oguz, U. ; Mchlaughlin,M. ;Garett, G.G. *Tetrahedron Lett*, **2002**, 43, 2873-2875.

[21] Malin, S.A.; Laskin, B.M.; Malin, A.S. Russ. J. Appl *Chem, **2007**,* 80, 2165-2168.

[22] Parous, J. ; Mealy, N. *J. Drugs future*, ***1994***, 19,111.

[23] Bouschi, F ; Camps, P. ; Franchini , M.C.; Torrero, D.; Ricci, A.; Sanchez, L. *Tetrahedron: Asymmetry*, ***2005***, 16, 3739-3745.

[24] Cahours, A. *Ann*, ***1859***, 109,10-34.

[25] Heintz, W. *Ann*, ***1862***, 122, 257-294.

[26] Tobie, W., Ayres, B. G., *J. Am. Chem. Soc.* ***1937***, 59, 947-950.

[27] *(a)* Corey, E. J.; Link .J. O., *Tetrahedron Lett.* ***1992***, 33, 3431-3434.
(b) Seebach, D. ; Naef.R. ; Calderari, G. *Tetrahedron*, ***1984***, 40,. 1313-1314.

[28] Dilber, S. P.; Dobric, S. L.; Juranic, Z. D.; Markovic, B. D.; Vladimirov, S. M.; Juranic, I.
O. *Molecules*, ***2008***, 13, 603-615.

[29] Kunz, H.; Larchen, H. G. *Tetrahedron Lett*, ***1987***, 28, 1873-1876.

[30] *(a)* Hell. C. *Eur. J. Inorg Chem*, ***1881***, 14, 891-893.

(b) Volhard, J., *Eur. J. Inorg. Chem*, ***1887***, 242, 141-163.
(c) Zelinesky, N. *Ber. Dtsh. Chem. Ges.* ***1887***, 20, 2026-2026.
[31] Watson, H. B. *Chem. Rev*, ***1930***, 7,173-201.
[32] Zhang, L. H.; Duan, J.; Xu, Y.; Dolbier, W. R. *Tetrahedron Lett.* ***1988***, 39, 9621-9622.
[33] ***(a)*** Stotter, P. L.; Hill, K. A., *Tetrahedron Lett*, ***1972***, 40, 4067-4070.
[34] Amanetoullah, A.O.; Chaabouni, M. M.; Baklouti, A., J. Fluorine Chem, 1997, 84, 149153.
[35] Robert, A.; Jagulin, Guinamant, S. *Tetrahedron*, ***1986***, 42, 2275-2281.
[36] Coppola, G. M. ; Schuster, H. F. Asymmetric synthesis: "Construction of chiral molecules using amino acids", Weily ; New York, ***1987***.
[37] larchveque, M.; Petit, Y. *Bull. Soc. Chim. Fr*, ***1989***, 1, 130-139.
[38] Tadashi, A.; Toshio, T., Mutsio, K. *Tetrahedron Lett*, ***2004***, 45, 1873-1876.
[39] Mahajan, R. P; Patil, U.K.; Patil, S. L. *India J.Chem.* ***2007***, 46, 1459-1465.
[40] Smita, R. W.; Harsh, K. G. *J. Chem. Pharm. Res*, ***2012***, 4, 2415-2421.
[41] Yoshioka, T.; Fujita, T.; Kanai, T.; Aizawa, Y.; Kurmumada, T.; Hasegawa, K.; Horikoshi, H. *J. Med. Chem.* ***1989***, 32, 421-428.
[42] Denis, R. St. L. ; Qi, G. ; Dedong, W. ; Michael, H. S.W. *Tetrahedron Lett.* **2004**,45, 19071910
[43] Molteni, G.; Buttero, P. D. Tetrahedron, **2005**, 61, 4983-987
[44] Oldo, K.; Hoydahl, E. A.; Hansen, T.V. *Tetrahedron Lett*, **2007**,48, 2097-2099
[45] Khan, S.A., Saleem, K., Khan, Z.: *Eur. J. Med. Chem*, **2008**, 43, 2257-2261.
[46] King, F. ; Clark-Lewis,J.; *J. Chem. Soc.* **1951**, 3379, 1953.
[47] Seada, M.; Fawzy, M.; El-Bez,I ; Abdel -Rahman, R. M. *Boll. Chim. Farm*, **1994**, 133, 381-388.
[48] Allouche, F.; Chabchoub, F.; Dahaoui, S.; Ben Hassine, B.; Salem, M. *Phys. Chem, News,* **2002**, 8,124-130.
[49] Allouche, F.; Chabchoub, F; Ben Hassine, B.; Salem, M. Heterocyclic Commun, **2011**, 10, 63-66.
[50] Schultz, A. G.; Napier, J. J.; Ravichandran, R, *J. Org Chem*, **1983**, *48*, 3408-3412.
[51] Richard , T. A. ; Srdanka ,T. K. *J. Org Chem*, **1978**, 43, 19-23.
[52] Minoru ,O., ; Kunio ,W. ;Hidehito, H. ; Asako ,T.; Tomoko, K.; Sachiko, S.; Jun, O.; Nobufusa, S. ; Toshihiko, F. ; Hiroyoshi, H.; Takashi, F. *J. Med. Chem.* **2000**, *43,* 3052-3066.
[53] Hans, M., Eberhard, B. ; Konrad ,V. E. *J. für Praktische Chemie*, **1939**,152, 237-266.
[54] Aoyama,T.; Takido,T.; Kodomari, M. *Tetrahedron Lett*, **2004**, 45, 1873-1876.
[55] Brychtova, K,; Slaba, B.; Placek, L.; Jampilek, J.; Raich, I.; Csollei, J. *Molecules* , **2009**, 14, 3019-3029.
[56] Schwenk, E.; Papa, D. *J. Am. Chem. Soc.* **1948**, *70*, 3626-3627.
[57] Ogata,Y.; Sugimoto,T. *J. Org. Chem.* ***1978***. 43, 19-23.
[58] (a) Berkovitz, P.T.; Long, R. A.; Dea, P.; Robins, R. K.; Matthews, T. R. *J. Med. Chem*, **1977**, 20,134-138.
(b) Ke, S. Y.; Qian, X. H.; Liu, F. Y. ; Wang, N , Yang,Q. ; Li, Z. *Eur. J. Med. Chem*, **2009**, 44, 2113-2121.
(c) Tka, N. ; Jegham, N. ; Ben Hassine. B.; H. C. R. *Chimie*, **2010**, 13, 1278-1283.
[59] Trepanier, D. L.; Eble, J. N.; Harris, G. H. *J. Med. Chem.* **1968**, 11, 357-360.
[60] Wouters, J.; Ooms, F.; Jegham, S.; Koenig, J. S.; George, P.; Durant, F. *Eur. J. Med. Chem*, **1997**, 32, 721-730.

[61] Arikan, N.; Sumengen, D.; B. Dulger, B. *Turk, J. Chem*, **2008**, 32, 147-155.
[62] Piccionello, A. P.; Pace, A.; Buscemi, S.; Vivona, N.; Giorgi, G. *Tetrahedron. Lett*. **2009**, 50, 1472-1474.
[63] Hussein, A. Q.; El-Abdelah, M. M.; Sabri, W. S. *J. Heterocylic, Chem*, **1984**, 21, 455.
[64] Berkowitz,P. T; Long, R. A.; Dea, P.; Robins, R. K.; Matthews, T. R. *J. Med Chem*, **1977**, 20 ,134-138.
[65] **(a)** Trepanier, D. L.; Eble, J. M.; Harris, G. H. *J. Med. Chem*, **1968**, 11, 357-360.
(b) Rodrigues,S. ; Olivato,P.R. ; Roberto Rittner,R. *Synthesis*, **2005**, 15, 2578-2582.
[66] **(a)** Vaughn, J. F.; Hitchcock, S. R. *Tetrahedron: Asymmetry*, **2004**, 15, 3449-3455.
(b) Burgeson, J. R.; Dore, D. D.; Standard, J. M.; Hitchcock, S. R., *Tetrahedron*, **2005**, 61, 10965-10974.
[67] Hitchcock, S. R.; Nora, G. P.; Casper, D. M.; Squire, M. D.; Maroules, C. D.; Ferrence, G. M.; Szczepura, L. F.; Standard, J. M., *Tetrahedron*, **2001**, 17, 9789-9798.
[68] Roussi, F.; Bonina, M.; Chiaroni, A.; Micouin, Riche, C.; Husson, H.P., *Tetrahedron Lett*, **1999**, 40, 3727-3730.
[69] Roussi, F.; Bonina, M.; Chiaroni, A.; Micouin, Riche, C.; Husson, H. P. *Tetrahedron Lett*. **1998**, 39, 8081-8084.
[70] Westphalt,G. ; Müller,T. *Advanced Synthesis & Catalysis*, **1978**, 320, 452-456.
[71] Ţînţaş, M. L; Diac, A. P; Soran, A; Terec, A.; Grosu, I.; Bogdan, E. *J. Mol. Struct.* **2014**, 1058,106-113.
[72] Piccionello,A.P.; Pace, A.; Buscemi, S.; Vivona, N.; Giorgi, G. *Tetrahedron. Lett*. **2009**, 50, 1472-1474.
[73] **(a)** Young, J. R.; DeVita, R. J. *Tetrahedron Lett*, **1998**, 39, 3931-3934.
I.) Chiou, S.; Shine, H. J. *J. Heteocyclic*, **1989**, 26, 125-128.

Experimental part

II. General methods

I-1 Purification of solvents and reagents

Anhydrous solvents are prepared by distillation in an argon atmosphere in the presence of drying agents.

Dichloromethane is distilled over calcium chloride ($CaCl_2$) and then over calcium hydride (CaH_2).

Methanol is distilled over magnesium in the presence of a catalytic amount of iodine.

Thionyl chloride ($SOCl_2$) is distilled under argon just before use.

I-2. Reactions

The reactions, in an anhydrous medium, were carried out under an argon atmosphere and in glassware dried in an oven just before use. The organic solutions were concentrated under reduced pressure using a rotary evaporator.

I-3 Preparation of the developer

The potassium permanganate developer is prepared by dissolving 3g of KMnO4 and 20g of K2CO3 in 300mL of distilled water in the presence of 5 mL of a 5% aqueous sodium hydroxide solution and 2 mL of glacial acetic acid.

I-4 Chromatography

Thin layer chromatography

Analytical thin-layer chromatography (TLC) was performed on ready-to-use silica plates (silica gel with Fluka Art 254 nm fluorescent indicator) and developed using a UV lamp (254 nm).

"Column chromatography

The products were purified by column chromatography using silica gel as the stationary phase.

I-5 Analytical methods and equipment

* Nuclear Magnetic Resonance (NMR)

Nuclear Magnetic Resonance spectra were performed in solution in deuterated chloroform ($CDCl_3$) using a BRUKER AC-300 (300MHz) for H NMR[1] and (75 MHz) for C NMR[13] .

The internal reference used is tetramethylsilane (TMS). The chemical shifts δ (ppm) are positive towards the low fields compared with TMS and the noted coupling constants are expressed in Hertz (Hz). To describe the multiplicity of signals, the following abbreviations were used **(Tables IV):**

Abbreviation	Multiplicity
S	Singulet
Sl	Large singlet
D	Doublet
T	Triplet
Q	Quadruplet
M	Multiplet
Dd	Split doublet

III. Operating modes

II-1 Preparation of optically active α-bromoacids from amino acids

Case of (2S, 3S)-2-bromo-3-methylpentanoic acid: ***53d***

In a 100 mL flask, 11g (84 mmol) of (*L*)-isoleucine and 35g of KBr (294 mmol, 3.5 equivalents) are dissolved in 20.7 mL of a 2M hydrobromic acid solution. To this solution, which was magnetically stirred and cooled to -13°C (and in which an argon bubble was maintained), 6.37g (92.35 mmoles) of powdered sodium nitrite was added in small portions. The solution turns brown with each addition. A second portion of salt is added only when the solution has decolourised and no more nitrogen is released. Under these conditions, the addition takes about 3 hours. Stirring at a temperature of between -15 and -10 is then maintained for a further 2 hours, followed by 3 hours at room temperature until the nitrogen has been released. Next, 3 x 50 mL of chloroform is added to the solution. The organic phases were combined, dried over MgSO4, filtered and concentrated under reduced pressure. The residue obtained was purified by chromatography on a silica column (eluent: cyclohexane/ethyl acetate, 70:30). Compound **53d** was obtained in 91% yield.

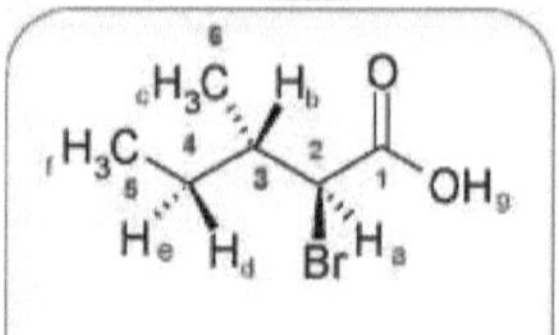

53d

Crude formula:C6H11O2Br.
M (g /mol): 195.
Developer: UV lamp.
Rdt=91%.
Appearance: Colourless oil.
NMR[1] H: (CDCl3, δ ppm); 0.93 (t, 3Hf, *J*= 7.2 Hz); 1.05 (d, 3Hc, *J*=6.6 Hz); 1.26-1.38 (m,1He); 1.70-1.79 (m, 1Hd); 2.00-2.09 (m, 1Hb); 4.12 (d, 1Ha, *J*=8.1Hz); 9.89 (sl, 1Hg).
NMR[13] **C** (75MHz, CDCl3): δ 10.10 (C5); 16.37 (Ce); 25.71 (C4); 37.60 (C3); 51.95 (C_2); 175.06 (CO).

*(2S)-2-bromopropanoic acid: **53a***

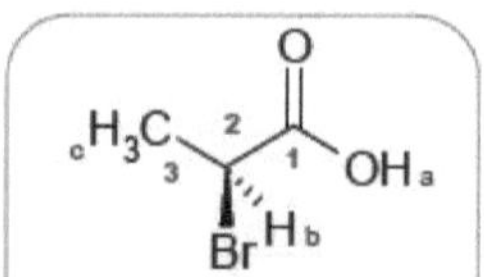

53a

Crude formula: C3H5O2Br.
M (g /mol): 153.
Developer: UV lamp.
Rdt=70%.
Appearance: Colourless oil.
NMR[1] H (CDCl3, δ ppm):1.83 (d, 3Hc, *J*= 6.9 Hz); 3.81 (q, 1Hb, *J*= 6.9 Hz); 9.44 (sl, 1H).
NMR[13] C (75MHz, CDCl3): δ 21, 47 (C3); 39, 52 (C_2); 175, 71 (CO).

*(2S)-2-bromo-3-methylbutanoic acid: **53 6***

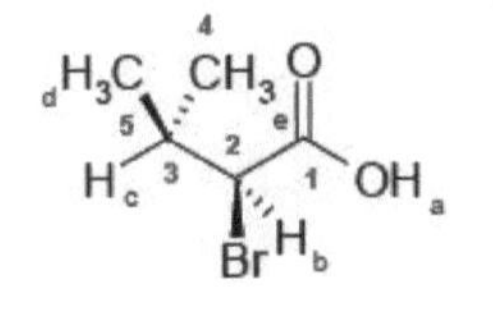

53b

Crude formula: $C_5H_9O_2Br$.
M (g /mol): 181.
Developer: UV lamp.
Rdt=98%.
Appearance: Yellowish oil
NMR[1] H (CDCl3, δ ppm); 1.08 (d, 3He, J = 6.9Hz); 1.12 (d, 3Hd, J = 6.6Hz); 2.13-2.29 (m, 1Hc); 4.08 (d, 1Hb, J=8.1Hz); 8.69 (sl, 1Ha).
NMR[13] C (75MHz, CDCl3): δ 18.22 (C4); 19.54 (C5); 31.46 (C3); 52.04 (C2); 174.65 (CO).
(2S)-2-bromo-4-methylpentanoic acid: ***53c***

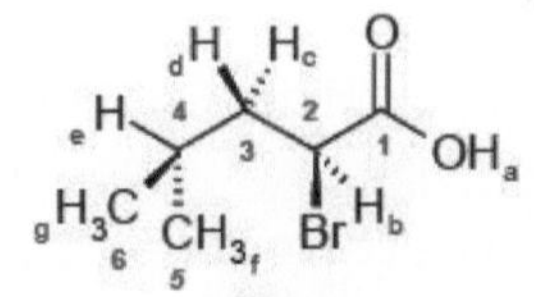

53c

Crude formula: $C_6H_{11}O_2Br$.
M (g /mol): 195.
Developer: UV lamp.
Rdt=90%.
Appearance: Yellowish oil.
Crude formula: $C_6H_{11}O_2Br$.
M (g /mol): 195.
NMR[1] H (CDCl3, δ ppm); 0.92 (d, 3Hf, J= 6.3Hz); 0.98 (d, 3Hg, J= 6.3Hz); 1.76-1.87 (m, 2Hc,d); 1.94 (m, 1He); 4.29 (t, 1Hb); 10.04 (sl, 1Ha).
NMR[13] C (75MHz,CDCl3): δ 21.59 (C5); 22.37 (C6); 26.34 (C4); 43.21 (C3); 44.08 (C2);176.31 (CO).
(2S)-2-bromo-3-phenylpropanoic acid: ***53e***

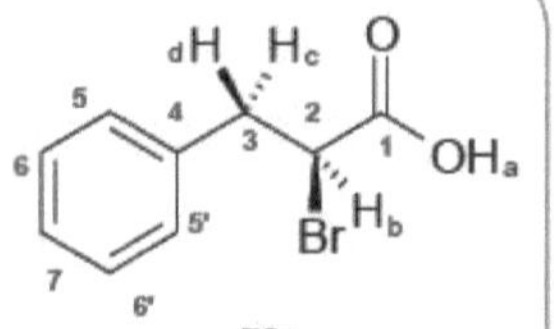

53e

Crude formula: C9H9O2Br.
M (g /mol): 229
Developer: UV lamp.
Rdt = 79%.
Appearance: Yellowish oil.
NMR[1] H (CDCl3, δ ppm); 3.34 (d, 1H); 3.39 (d, 1H_d); 4.25 (t, 1H); 7.23-7.42 (m, 5Ha_{rom}); 10.24 (sl, 1Ha).
NMR[13] C (75MHz, CDCl3) δ: 33.58 (C3); 60.09 (C_2); 126.62-143.68 (C4, C5, C6, C7) and 177.98 (CO).

*(2S)-2-bromo-4-methylpentanoic acids: **53f***

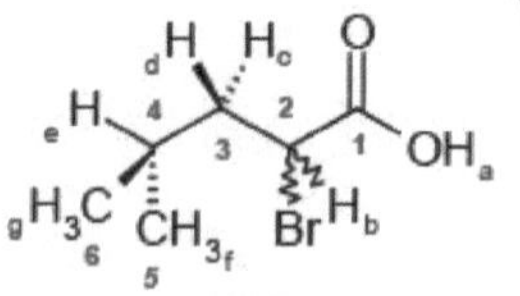

53f*

Gross formula: C H O6112 Br.
M (g /mol): 195.
Developer: UV lamp.
Rdt=93%.
Appearance: Yellowish oil.
NMR[1] H (CDCl3, δ ppm); 0.92 (d, 3Hf, J= 6.3Hz); 0.98 (d, 3Hg, J= 6.3Hz); 1.76-1.87 (m, 2Hc,d); 1.94 (m, 1He); 4.29 (t, 1Hb); 10.04 (s, 1Ha).
NMR[13] C (75MHz, CDCl3): δ 21.59 (C6); 22.37 (C5); 26.34 (C4); 43.21 (C3); 44.08 (C2); 176.31 (CO).

II-2 Synthesis of optically active α-bromoesters from a-bromo acids

Case of methyl 2-bromo-4-methylpentanoate: 93f

2-bromo-4-methylpentanoic acid (16 mmol) was dissolved in freshly distilled thionyl chloride (19 mmol, 1.2 equivalents) in a 100 mL flask fitted with a reflux vessel, followed by the addition of three drops of anhydrous DMF. The mixture was refluxed overnight under an argon atmosphere. Excess thionyl chloride was evaporated. Anhydrous methanol was then added in excess at low temperature (0°C) and two equivalents of triethylamine. The reaction mixture was stirred overnight and then concentrated in vacuo. The crude reaction product was purified by column chromatography [silica / (AcOEt /cyclohexane (30: 70)]. Compound **93f*** was obtained in 80% yield.

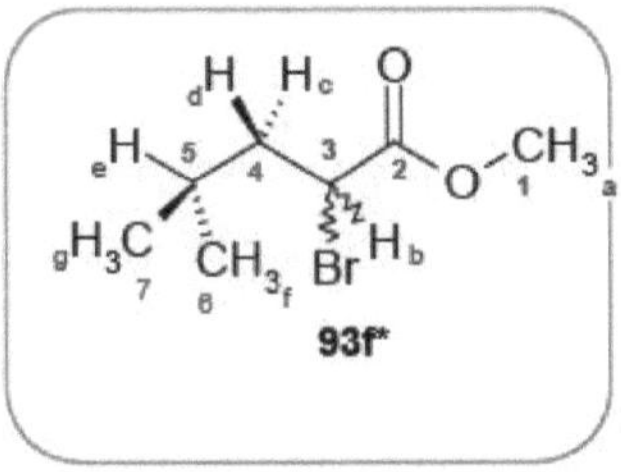

Gross formula: $C_7H_{13}O_2Br$.
M (g /mol): 209.
Developer: UV lamp.
Rdt=80%.
Appearance: Yellowish oil.
NMR[1] H ($CDCl_3$, δ ppm); 0.896 (d, 3Hf, J= 6.3Hz); 0.93 (d, 3Hg, J= 6.3Hz); 1.73-1.80 (m, 2Hc,d); 1.88-1.93 (m, 1He); 4.28 (t, 1Hb); 3.77 (s, 3Ha).
NMR[13] C (75MHz, $CDCl_3$): δ 21.03 (C_7); 21.82(C_6); 25.83(C_5); 43.01 (C_4); 43.82 (C_3); 52.37 (C_1); 170.06 (CO).

Case of methyl (2S)-2-bromo-3-methylbutanoate: ***936***

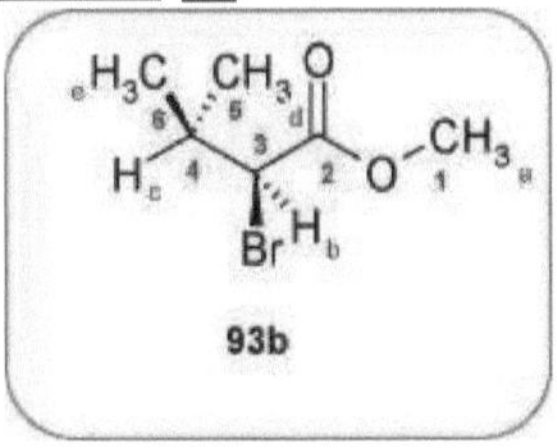

Crude formula: $C_6H_{11}O_2Br$.
M (g /mol): 195.
Developer: UV lamp.
Rdt= 85%.
Appearance: Yellowish oil.
NMR[1] H ($CDCl_3$, δ ppm); 1.03 (d, 3Hd, J = 6Hz); 1.11 (d, 3He, J = 6.3Hz); 2.16-2.28 (m, 1Hc); 3.79 (s, 3Ha); 4.06 (d, 1Hb, J=7.8Hz).
NMR[13] C (75MHz, $CDCl_3$): δ 19.36 (C_5); 19.50 (C_e); 31.85 (C_3); 52.24 (C_1); 169.47 (CO).

Case of methyl (2S,3S)-2-bromo-3-methylpentanoate: ***93d***

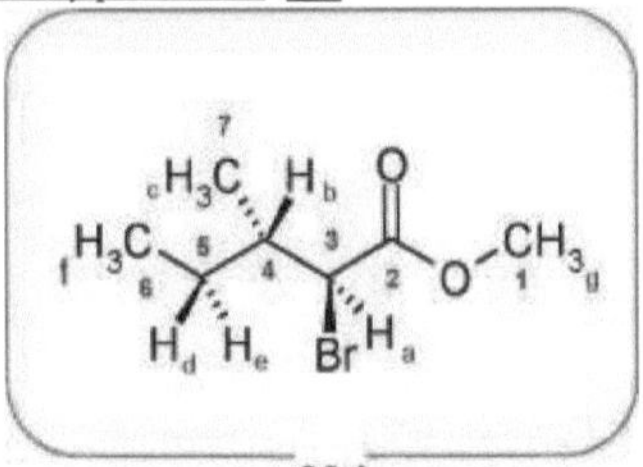

93d

Gross formula: $C_7H_{13}O_2Br$.
M (g /mol): 209.

Developer: UV lamp.
Rdt= 90%.
Appearance: Colourless oil.
NMR[1] H: (CDCl3, δ ppm); 0.90 (t, 3Hf, J= 7.5 Hz); 0.97 (d, 3Hc, J= 6.9 Hz); 1.24-1.33 (m, 1He), 1.69-1.76 (m, 1Hd); 1.99-2.05 (m, 1Hb); 4.08 (d, 1Ha, J=8.4Hz); 3.76 (s, 3Hg).
NMR[13] **C** (75MHz, CDCl3): δ 10.07 (C6); 15.68 (C_7); 25.78(C5); 37.83(C_4); 52.21 (C3); 53.28 (C1); 169.48 (CO).

Case of methyl (2S)-2-bromo-4-methylpentanoate: ***93c***

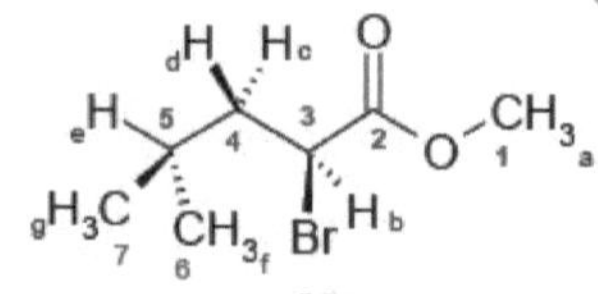

93c

M (g /mol): 209.
Developer: UV lamp.
Rdt= 83%.
Appearance: Yellowish oil.
NMR[1] H (CDCl3, δ ppm); 0.896 (d, 3Hf, *J*= 6.3Hz); 0.93 (d, 3Hg, *J*= 6.3Hz); 1.73-1.80 (m, 2Hc,d); 1.88-1.93 (m, 1He); 3.77 (s, 3Ha); 4.28 (t, 1Hb).
NMR[13] C (75MHz, CDCl3): δ 21.03 (C6); 21.82 (C7); 25.83 (C5); 43.01 (C4); 43.82 (C3); 52.37 (C2); 170.06 (CO).

Case of methyl (2S)-2-bromo-3-phenylpropanoate: ***93e***

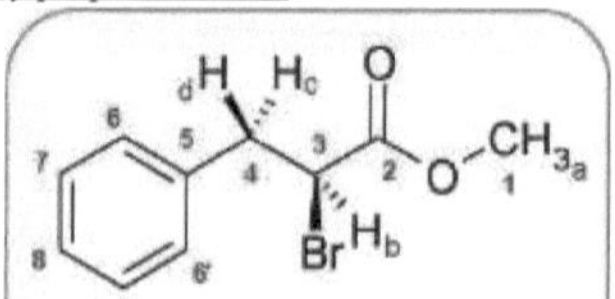

93e

Crude formula: C10H10O2Br.
M (g /mol): 242.
Developer: UV lamp.
Rdt = 76%.
Appearance: Yellowish oil.
NMR[1] H (CDCl3, δ ppm); 3.31 (d, 1H); 3.46 (d, 1Hd); 3.74 (s, 3Ha); 4.45 (t, 1H); 7.23-7.42 (m,5H).
NMR[13] C (75MHz, CDCl3) δ: 40.60 (C4) ; 44.63 (C3) ; 52.39 (CI) ; 126.87(C_8); 128.19 (C6,6') ; 128.42 (C7,7') ; 136.25 (C5) ; 169.16 (CO).

Case of methyl (2S)-2-bromopropanoate: ***93a***

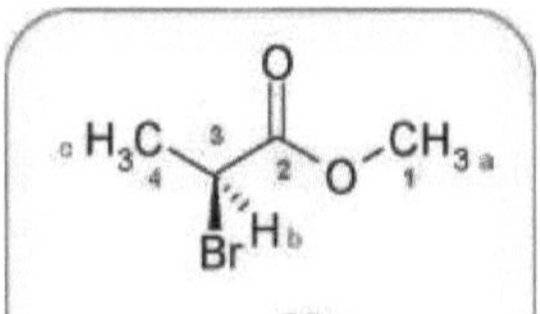

93a

Crude formula: C4H7O2Br.
M (g /mol): 167.
Developer: UV lamp.
Rdt= 78%.
Appearance: Colourless oil.
NMR[1] H (CDCl3, δ ppm):1.83 (d, 3Hc, J= 6.9 Hz); 3.81 (q, 1Hb J= 6.9 Hz); 3.83 (s, 3Ha).
NMR[13] C (75MHz, CDCl3): δ 21, 47 (C3); 39, 52 (C3); 50.38 (C1); 170.22 (CO).

II-3 Preparation of 6-alkyl-3-(1'-benzenesulfonylpyrrolidin-2'-yl)-4,6-dihydro-1,2,4-oxadiazin-5-ones

Case of (2S, 6R) and (2S, 6S) 6-methyl-3-(1'-benzenesulfonylpyrrolidin-2'-yl)- 4,6-dihydro-1,2,4-oxadiazin-5-one: ***130a***

Dissolve 0.13 g (7.8 mmoles) of methyl (*2S*)-2-bromopropanoate and one equivalent of (*2S*)-1-bzenesulfonylpyrrolidin-2-carboxamidoxime in anhydrous dichloromethane in a 25 mL flask under an atmosphere of argon at room temperature. Two equivalents of NaH were then added. The reaction medium was stirred at room temperature for two hours. The reaction was monitored by TLC. The solvent was then evaporated. The crude reaction mixture was extracted three times with dichloromethane. The organic phase was washed with a saturated solution of NaCl and dried over MgSO4. After evaporation of the solvent, the residue was purified by silica column chromatography using the cyclohexane/ethyl acetate (50:50) system, then recrystallised in [chloroform/petroleum ether] to give a mixture of two diastereoisomers of compound **130a**.

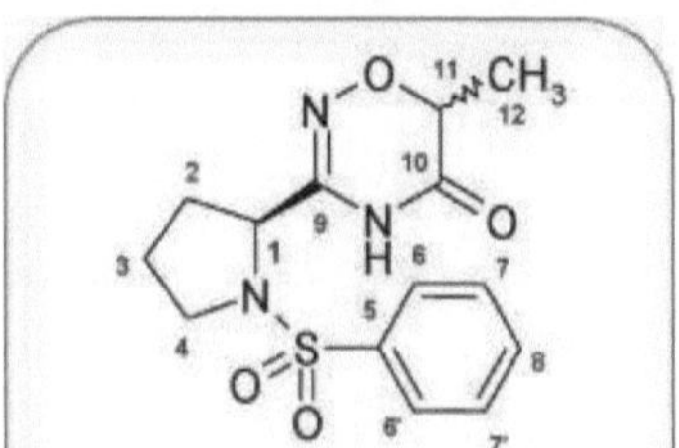

130a

Gross formula: C14H17O4SN3 .
M (g /mol): 307.
Developer: UV lamp.
Rdt = 78%.
Appearance: White solid.
Crystallisation solvent: chloroform / petroleum ether.
NMR[1] H (300MHz , CDCl3): δ 1.53(d, 3Hn , *J* = 6.9Hz); 1.67 (m, 2H3,3'); 2.06 (m, IH4); 2.52 (m, 1H4'); 3.25 (m, 1H2); 3.61 (m, 1H2');4.25 (m, 2H11,1); 7.61 (t, 2H7,7'); 7.68 (d, 1H8); 7.86

(d, 2H6,6'), 8.84 NH.

NMR[13] C (75MHz, CDCl3), δ12.37 (c12); 24.45 (c3); 26.91 (c4); 49.81 (c2); 58.58 (ci); 72.78 (c11); 127.87 (c6,6'); 129.57 (C7,7'); 133.77 (c8); 135.27(c5); 151.36 (c9); 166.91(c10).

*Case of (2S, 6R) and (2S, 6S) 6-isopropyi-3-(1'-benzenesuifonyipyrroiidin-2'-yl)-4,6-dihydro1,2,4-oxadiazin-5-one:**1306***

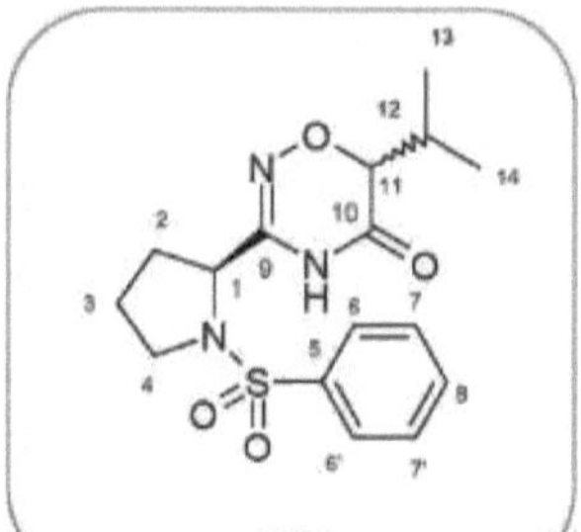

130b

Gross formula: C16H21O4SN3 .

M (g /mol): 335.

Developer: UV lamp.

Rdt = 80%.

Appearance: White solid.

Crystallisation solvent: chloroform / petroleum ether.

NMR[1] H (300MHz, CDCl3): δ 1.06-1.149 (m, 6H13,14); 1.61-1.71 (m, 2H3,3'); 1.93 (m,1H_n); 2.35-2.41 (m, 2H4,4'); 3.26 (m, 1H2); 3.57 (m, 1H2'); 3.95 (d, 1Hh, *J* =5.7Hz); 4.27 (dd, 1Ha, *J* = 7.5Hz, *J* = 3.6Hz); 7.63 (t, 2H7.7'); 7.67 (d, 1H8); 7.87 (d, 2H6.6'); 8.89 NH.

NMR[13] C (75MHz, CDCl3), δ16.72 (c13); 18.25 (c14); 26.96 (c12); 28.95 (c4); 49.25 (c2); 58.11 (c1); 80.36 (c11); 127.01 (c6); 127.33 (C6'); 128.83 (c7); 129.03 (C7'); 133.20 (c8); 135.04 (c5); 150.18 (c9); 165.81 (c10).

*Case of (2S, 6R) and (2S, 6R) 6-isobutyi-3-(1'-benzenesulf0nyipytrolidm-2'-yl)-4,6-dihydro-1,2,4- oxadiazin-5-one:**130c***

12

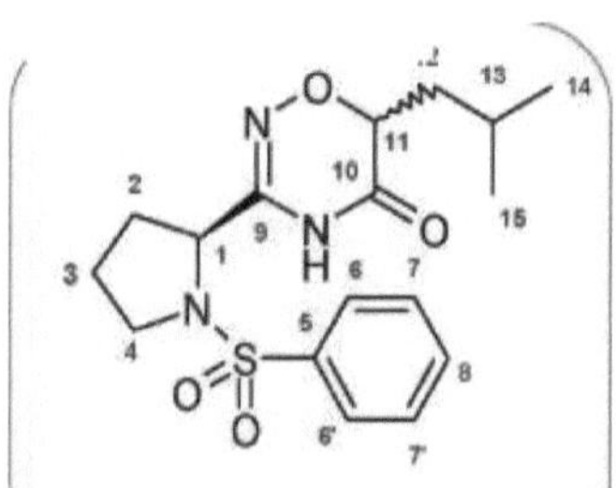

130c

Gross formula: C17H23O4SN3 .

M (g /mol): 349.

Developer: UV lamp

Rdt = 83%.

Appearance: White solid.

Crystallisation solvent: chloroform / petroleum ether.

NMR[1] H (300MHz, CDCl3): δ 0.98 (m, 6H14.15); 1.65-1.74 (m, 3H3,3',13); 1.78 (m, 2‰); 1.97 (m, 1H4); 2.29 (m, 1H4'); 3.26 (m, 1H2); 3.59-3.62 (m, 1H2'), 4.19- 4.28 (m, H11,1); 7.59 (t, 2H7,7'); 7.65 (d, 1H8); 7.84 (d, 2H6,6'); 8.92 NH.

NMR[13] C (75MHz, CDCl3), δ 21.73 (C14); 23.09 (C15); 24.50 (C13); 26.89(C3); 29.66 (C12); 36.40 (C4); 48.89 (C2); 58.61(C1); 74.26 (C11); 127.49 (C7,7'); 129.25 (C6,6'); 132.73 (C8); 135.52 (C5); 151.30 (C9); 167.090 (C10).

Case of (2'S, 6R) and (2S, 6S) 6-sec-butyi-3-(1'-benzenesu[fonylpyrro[idin-2'-yl)-4,6-dihydro-1,2,4-oҳadiazin- 5-one: ***130d***

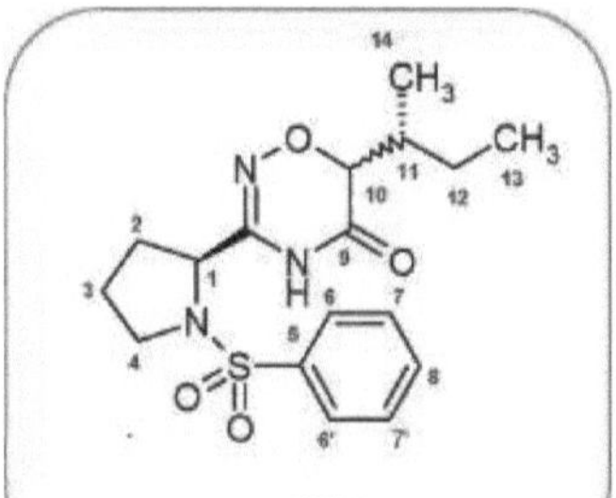

130d

Gross formula: C17H23O4SN3 .

M (g /mol): 349.

Developer: UV lamp

Crystallisation solvent: chloroform / petroleum ether.

Rdt = 89%.

Appearance: White solid.

NMR[1] H (300 MHz , CDCl3): δ 0.97 (t, 3H$_n$); 01.06 (d, 3‰), 1.39(m, 1H), 1.57 (m, 3H), 1.96(m, 1H), 2.12 (m, 1H); 2.28 (m, 1H); 3.16 (m, 1H2'); 3.50 (m, 1H2), 4.06 (d, 1H10); 4.08 (t, 1H1); 7.63 (m, 2H7,7'); 7.67 (d,1H8); 7.87 (d, 2H6,6'), 8.85 NH.

NMR[13] C (75MHz, CDCl3), δ 12.08(C14);1.43(C15); 24.89 (C13); 25.68(C3); 29.23(C$_n$); 33.96(C4); 49.75(C2); 58.95(C1);79.04(C11); 128.95(C6,6'); 129.32(C7,7'); 133.70(C8); 135.51(C5); 150.72 (C9); 166.43(C10).

Case of (2S, 6R) and (2S, 6R) 6-benzyl-3-(1'-benzenesulfonylpyrrolidin-2'-yl)-4,6-dihydro-1,2,4-oҳadiazin-5-one: ***130e***

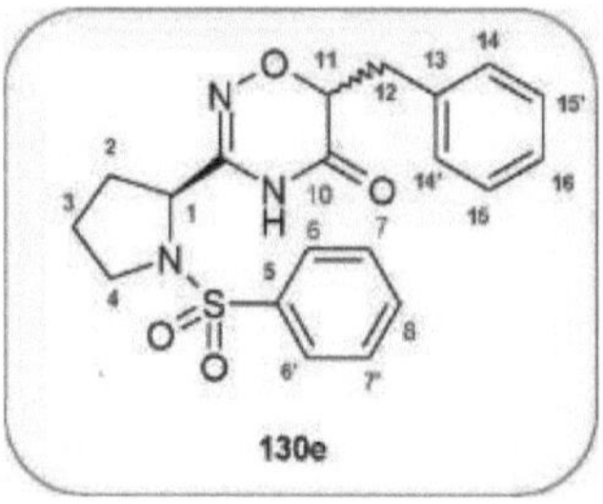

130e

Gross formula: C20H21O4N3S.

M (g /mol): 383.

Developer: UV lamp

Crystallisation solvent: chloroform / petroleum ether.
Rdt = 75%.
Appearance: White solid.
NMR[1] H (300 MHz , CDCl3): δ 1.64 (m, 2H); 1.92(m, 1H); 2.30 (m, 1H); 3.19 (m, 2H); 3.34(m, 1H2'); 3.56 (m, 1H2); 4.25 (m, 1H11); 4.37 (m, 1H1); 7.32-7.75 (m, 10H), 8.78 NH.
NMR[13] C (75MHz, CDCl3), δ 24.83(C3); 29.30 (C12); 34.64(C4); 50.17(C2); 58.92(C1); 78.52(C11); 126.89; 127.54; 127.85; 128.47; 128.99; 129.58; 132.75; 135.29; 151.37 (C9); 166.37(C10).

General conclusion and outlook

In the search for new chiral heterocycles with potential biological activity, we have synthesised a series of five new chiral 1,2,4-oxadiazin-5-ones.

The synthesis strategy adopted in the course of this work consists in first preparing a series of optically active a-bromo acids from amino acids in yields ranging from 70% to 98%.

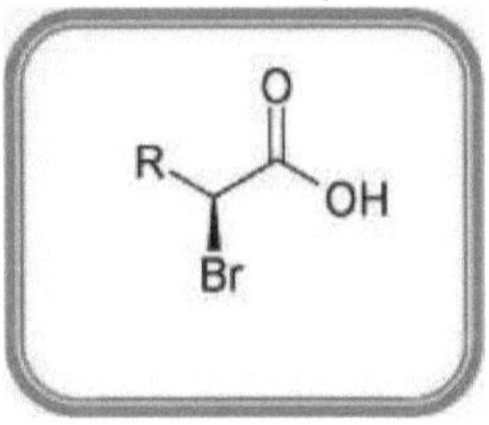

a: R=Me; **b**: R= i-Pr; **c**: R= i-Bu; d: R= Sec-Bu; e: R= Bn;* f: R= i-Bu

In a second step, we transformed the a-bromo acids into the corresponding a-bromoesters using acid chloride. The chemical yields of the a-bromoesters obtained were between 76% and 90%.

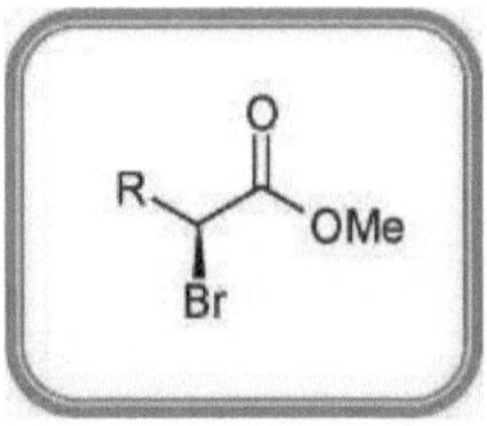

a: R=Me; **b**: R= i-Pr; **c**: R= i-Bu; d: R= Sec-Bu; e: R= Bn;* f: R= i-Bu

Condensation of the a-bromoesters obtained with 1-benzenesulphonylpyrrolidin-2-carboxamidoxime enabled the synthesis of a series of five chiral substituted 1,2,4-oxadiazin-5-ones-6 in good chemical yields of between 75% and 89%.

The spectroscopic study showed that these products were obtained as a mixture of two diastereoisomers.

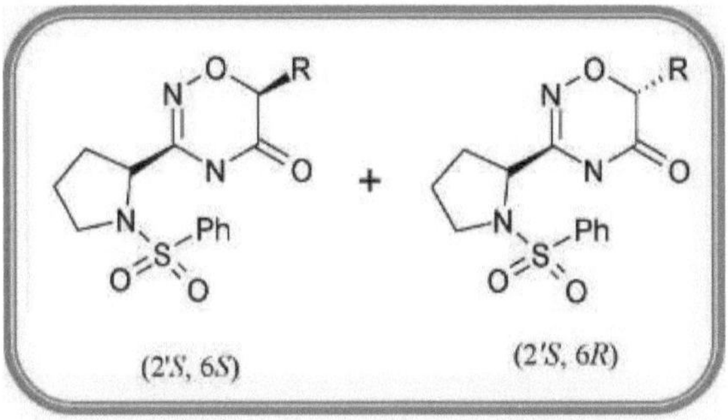

a : R=Me ; **b** : R= i-Pr ; **c** : R= i-Bu ; **d** : R= Sec-Bu ; **e** : R= Bn .

We hope to build on this work in the future:

4- Use the a-bromoesters prepared in other reactions to prepare biologically active heterocycles.

4- Try to separate the two diastereomers obtained.

4- Test the biological activity of the products synthesised.

MIX
Papier aus verantwortungsvollen Quellen
Paper from responsible sources
FSC® C105338

Printed by Books on Demand GmbH, Norderstedt / Germany